THE TUNNELS

www.**penguin**.co.uk

THE TUNNELS

THE UNTOLD STORY OF THE ESCAPES
UNDER THE BERLIN WALL

Greg Mitchell

CORGI BOOKS

TRANSWORLD PUBLISHERS
61–63 Uxbridge Road, London W5 5SA
www.penguin.co.uk

Transworld is part of the Penguin Random House group of companies
whose addresses can be found at global.penguinrandomhouse.com

Penguin
Random House
UK

First published in Great Britain in 2016 by Bantam Press
an imprint of Transworld Publishers
Corgi edition published 2017

Frontispiece: Bernauer Strasse in the early 1960s. Photograph by Günter Zint.

A CIP catalogue record for this book
is available from the British Library.

ISBN
9780552172042

Typeset in Walbaum
Printed and bound by Clays Ltd, Bungay, Suffolk.

Penguin Random House is committed to a sustainable
future for our business, our readers and our planet. This book
is made from Forest Stewardship Council® certified paper.

MIX
Paper from
responsible sources
FSC® C018179

1 3 5 7 9 10 8 6 4 2

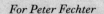

For Peter Fechter

He only earns his freedom who daily conquers anew.

—Goethe, *Faust*

CONTENTS

A NOTE TO READERS

The Tunnels adheres strictly to the historical record and reflections by participants and witnesses. It incorporates no invented dialogue. Re-created scenes are not imagined but based in most cases on accounts of two or more participants. Unless otherwise attributed, anything between quotation marks is either actual dialogue (as recalled by a witness, often in an interview with the author) or from a memoir or other book, letter, oral history, court record, interrogation, White House transcript, or other document cited in the Notes. In some quotations I have corrected syntax or punctuation. All of the names are real. Addresses in Berlin, where street names are combined with "strasse" as one word (e.g., Schönholzerstrasse), are rendered here for clarity with "Strasse" or "Platz" or "Allee" as separate words.

To an extent that surprised even the author, nearly all of the central events and episodes in this narrative (and surely the most exciting sections) are based on lengthy original interviews with nearly all of the key tunnelers, and several of the couriers and escapees; hundreds of pages of never-before-seen documents from the Stasi archives; and recently declassified State Department and CIA files.

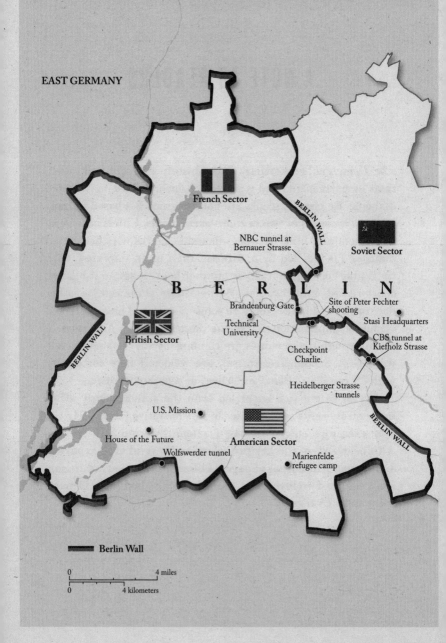

EAST GERMANY

French Sector

Soviet Sector

NBC tunnel at
Bernauer Strasse

B E R L I N

Brandenburg Gate

Site of Peter Fechter
shooting

Technical
University

Stasi Headquarters

British Sector

CBS tunnel at
Kiefholz Strasse

Checkpoint
Charlie

Heidelberger Strasse
tunnels

U.S. Mission

House of the Future

American Sector

BERLIN WALL

Wolfswerder tunnel

Marienfelde
refugee camp

━━━ Berlin Wall

0   4 miles
0   4 kilometers

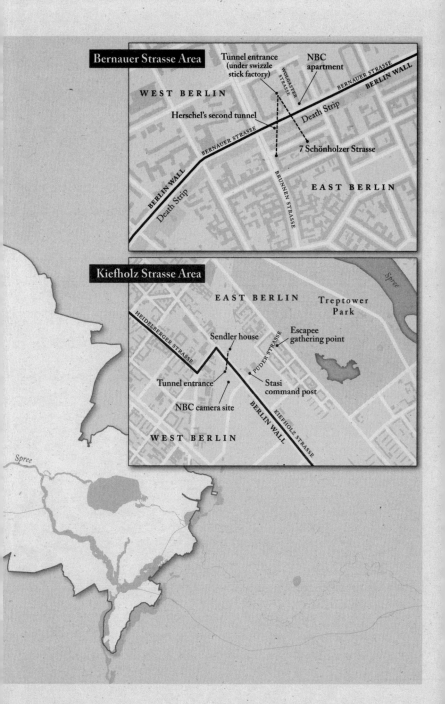

Bernauer Strasse Area

WEST BERLIN

Tunnel entrance
(under swizzle
stick factory)

NBC
apartment

WOLGASTER STRASSE

BERNAUER STRASSE

BERLIN WALL

Herschel's second tunnel

Death Strip

BERNAUER STRASSE

BERLIN WALL

Death Strip

7 Schönholzer Strasse

BRUNNEN STRASSE

EAST BERLIN

Kiefholz Strasse Area

EAST BERLIN

Treptower
Park

Spree

HEIDELBERGER STRASSE

Sendler house

PUDER STRASSE

Escapee
gathering point

Tunnel entrance

Stasi
command post

NBC camera site

BERLIN WALL

KIEFHOLZ STRASSE

WEST BERLIN

Spree

1

The Cyclist

FEBRUARY–MARCH 1962

Harry Seidel loved action, speed, risk. He found them all in bicycle racing. Harry might have been an Olympic champion—still could be, probably—if he changed his attitude, for at twenty-three he remained in his leg-churning prime. But that wasn't Harry. Once he set his mind on something he went full bore, and now he wasn't chasing the next turn, other racers, or a finish line. Just months ago he had competed before thousands of cheering fans in raucous arenas. His picture appeared in newspapers. Children might call out to the lean, dark-haired sports hero when they recognized him cycling on the streets of Berlin. Now he toiled nearly alone. No one cheered, even if he deserved it for victories far beyond any of his racing exploits. That would be too dangerous.

Since the emergence of the new barrier dividing Berlin on August 13, 1961, Harry's wife, Rotraut, had worried about him. Whenever he set off on one of his secret missions she wondered if he would fail to come home, perhaps forever. Friends called Harry a *draufgänger*—a daredevil. They urged him to quit his death-defying deeds, return to cycling, and open that newspaper kiosk he coveted, but they might as well have been shouting into a wintry wind off the River Spree. In just the first months after the Wall arrived, Seidel had led his wife and son, and more than two dozen others, across the nearly impenetrable border to the West. And in Harry's mind there were still countless others (that is, nearly anyone in the East) to rescue.

Seidel had drawn only praise from the state during his cycling career, which had culminated in several East Berlin titles and two medals at the 1959 East German championships. Barely out of his teens, he quit his job as an electrician when the state began paying him to compete full-time. Even as he was being extolled in propaganda organs, Harry revealed himself as insufficiently patriotic when, unlike many others on the national team, he refused to ingest steroids to enhance his performance. He also failed to join the ruling Communist Party. This cost him any chance to make the country's 1960 Olympic team, and his government stipend was canceled.

Now, in early 1962, his reputation in East German secret police files as an escape helper matched his fame as a cyclist. The trade had not come without cost.

Seidel's first escape had been his own. Just hours after the wire and concrete barrier materialized to brutally divide Berlin on the morning of August 13, Seidel left the apartment he shared with his wife, son, and mother-in-law in the Prenzlauer Berg district to explore the border by bicycle. South of the city center he found a spot where the barbed wire was low. With guards distracted by protesters, he shouldered his bike and leaped over the wire. It was a test more than anything. He figured he could return to the East just as easily—which he did, a few hours later, passing through a checkpoint. (It was still no problem going in *that* direction.) Harry being Harry, he felt confident he could jump the border again in the hours ahead. He wasn't eager to abandon Rotraut and baby Andre, but he didn't want to lose the newspaper delivery job he held in the West. Even if he did get stuck across the barrier he would surely find a way to get his family, including his mother, out soon.

Later that day Harry considered another vault to the West, but it looked like the border guards were tightening their controls. Just after dark he wrapped his passport in plastic and dove into the Spree to swim the more than two hundred yards to the West. Coming up for air he nearly head-butted an East Berlin police boat. Treading water, he finally heard one of the cops say, "Let's go, nothing to see here." After they left he swam the rest of the way to the shore.

While Seidel pondered how to rescue his family, one of Rotraut's brothers tried to get them out using West German passports bearing photos that resembled them. When that brother attempted to smuggle the fake IDs through a check-point, they did not pass muster. Harry's mother and mother-in-law were arrested. His wife remained free only because she had a baby to care for. Harry, enraged, vowed to retrieve his mother when she emerged from prison—and to spring his wife and son immediately.

After another bicycle tour, this time along the Western side of the Wall, he determined that the safest place for a breakout was along Kiefholz Strasse, near Treptower, one of the city's largest parks. There was nothing but barbed wire—no fencing or concrete—at the border there, and plenty of trees and bushes in the American-occupied zone for cover. To provide a blanket of darkness he shot out a couple of spotlights with an air rifle.

On the evening of September 3, 1961, three weeks after the coming of the Wall, Rotraut, slender and blue-eyed, received an unexpected phone call at her apartment. Harry, calling from a café in the East, announced that he would pick her up in an hour. Rotraut, whose family had emigrated from Poland, was as anti-Communist as her husband and had been considering ways to escape on her own, so the invitation from Harry was most welcome. When he arrived he told her to dress in black, give their baby part of a sleeping pill, and follow him. Soon they were penetrating the underbrush along Kiefholz Strasse, where Harry had already cut the barbed wire. He crawled through, then stood and lifted the top wire. Rotraut passed him the baby and stepped into the West. Then with Harry she ran like hell to his Ford Taunus. Minutes later the three Seidels were relaxing in Harry's apartment in the Schöneberg district.

The ending was not so happy for two of Rotraut's brothers, who were arrested on charges that they knew about or assisted the escape.

Few in East Berlin imagined that any sort of wall—or "anti-Fascist protection barrier," as East German leader Walter Ulbricht dubbed it (proving he had read his Orwell)—could last for years. But Harry Seidel was not among the optimistic. He believed the vast, ugly scar and police state were meant to be permanent. And what could the West do about it? Berlin was a fractured island floating precariously in the middle of the Communist state, one hundred miles from West Germany. Harry Seidel sensed that his adventures at the border had barely begun. For one thing, he still had to rescue his mother.

AFTER years of shortages and rationing, East Berliners liked to quip that even when they could afford to buy apples and potatoes they often found worms in them—and "they charge more with the worms." Another bitter joke: "Did you know that Adam and Eve were actually East Germans? They had no clothes, they had to share an apple, and they were led to believe that they lived in a paradise."

Since shortly after World War II, a wavy line on the map had separated the two German states, even before they took the names German Democratic Republic (GDR) and Federal Republic of Germany (FRG). West Germany was divided into sectors occupied by the Americans, the French, and the British. The Soviet-dominated GDR was Germany's junior half, in landmass, population, and increasingly, economic performance. In 1955, with its economy booming and jobs plentiful, West Germany achieved full sovereignty, even as the three occupying forces remained. The Communists in the East, meanwhile, scrambled to stem an embarrassing refugee crisis. From the late 1940s to 1961 some 2.8 million East Germans fled to the West.

Most of this human tide, nearly 20 percent of the East German population and a high concentration of its skilled workers and professionals, exited via Berlin. GDR soldiers tightly policed the national boundary, but the sector border at Berlin, deep inside East Germany, remained porous. Levels of security varied wildly where the city's four sectors met. Berlin remained, in most ways, one city, with interconnected telephone service, subway, train, tram, and bus lines. As many as sixty thousand East Berliners with official passes—teachers, doctors, engineers, lawyers, technicians, students—crossed into the West every weekday to work or attend classes at the Technical University or the Free University. They were known as *grenzgänger*—border crossers. Many never returned. By 1961, West Berlin's population of 2.2 million doubled that of the Eastern sector.

The Soviets grew alarmed. Premier Nikita Khrushchev considered West Berlin "a bone in my throat," even as he also likened it to testicles he could squeeze whenever he wanted the West to scream. Khruschchev had issued an ultimatum in November 1958 giving the three Western nations six months to agree to make West Berlin a "free," demilitarized zone, and then withdraw. The Allies rejected this. They held that the unnatural division of the city had to end in free elections in every sector and, ultimately, in reunification. Khrushchev backed down for the moment. Running for president in 1960, John F. Kennedy predicted Berlin would continue to be a "test of our nerve and our will."

The first Kennedy–Khrushchev summit took place in early June, 1961, in Vienna. The sixty-seven-year-old Soviet leader opened by calling Berlin "the most dangerous place in the world." Testing the inexperienced JFK, he threatened to finally sign a long-promised "peace treaty" with East Germany, ending the four-power agreements on sharing Berlin. The East Germans would thereby gain control of all Western access to the city via air, rail, and autobahn. Again, the three Western nations rejected the idea. Yet a fumbling, intimidated Kennedy hinted that the United States now accepted the semipermanent division of Berlin, which only emboldened Khrushchev.

As the summit ended, Kennedy privately called it "the worst

thing in my life. He savaged me." JFK told aides there was little America could do for the East Berliners—the sole goal now was to defend the interests of those already in the West. He assured a top aide, "God knows I'm not an isolationist, but it seems particularly stupid to risk killing a million Americans over an argument about access rights on an autobahn . . . or because the Germans want Germany reunified." After all, he added, "We didn't cause the disunity in Germany."

In a July 25, 1961, speech, Kennedy declared that the United States was not looking for another confrontation on Berlin. Still, in light of the Soviets' growing belligerence there, JFK ordered a military buildup. "We seek peace," Kennedy announced, "but we shall not surrender." West Berliners focused on another element of the speech: Kennedy seemed to suggest that while America would remain a strong defender of West Germany, it would let the Communists do pretty much whatever they wanted in the East. Amid the growing tensions, the number of East Germans arriving at West Berlin's refugee center, a colony of twenty-five buildings at Marienfelde, spiked. The rate had averaged 19,000 a month in 1961; this more than doubled in early August. East Germans had never been allowed to participate in free elections but they were voting with their feet.

Walter Ulbricht, the sixty-eight-year-old East German leader with a Lenin goatee, had seen enough. With Khrushchev's blessing, he had weeks earlier ordered the stockpiling of massive quantities of barbed wire, fencing, and concrete blocks, his fantasy of a permanent barrier encircling West Berlin suddenly about to come to life. Somehow, despite their vast investment in intelligence operations in Berlin, the Americans knew little about any of this. President Kennedy's daily CIA briefings mentioned nothing.

Not that it likely mattered. American leaders were profoundly ambivalent about the prospect of any sealing of the border. Ulbricht took heart from a well-publicized July 30 television interview with J. William Fulbright, an influential Democratic U.S. senator. Asked whether the Communists might reduce tensions by barring refugee flight, Fulbright answered, "Next week, if they chose to close their borders, they could without violating any treaty. I don't understand why the East Germans don't close their border. . . . I think they have a right to close it at any time." West German media and American diplomats in Bonn, the capital, excoriated Fulbright. Some called him "Fulbricht."

President Kennedy said nothing in public. But at the White House he told an adviser, "Khrushchev is losing East Germany. He cannot let that happen. If East Germany goes, so will Poland and all of Eastern Europe. He will have to do something to stop the flow of refugees. Perhaps a wall. And we won't be able to prevent it." Khrushchev, meanwhile, assured Ulbricht, "When the border is closed, the Americans and West Germans will be happy." He claimed that the American ambassador to Moscow had told him the increasing intensity of the refugee flight was "causing the West Germans a lot of trouble. So when we institute these controls, everyone will be satisfied." Ulbricht assigned his security chief, Erich Honecker, to make sure the operation succeeded.

Just after midnight on August 13 the first barbed wire was unrolled along major boulevards at the border, the first step in sealing off the ninety-six-mile circumference of West Berlin. Thousands of Soviet troops stood in reserve in case demonstrators in the West tried to stop it. Khrushchev had wisely advised Ulbricht to make sure the wire did not extend even one inch across the border.

When Secretary of State Dean Rusk heard the news later that morning, he ordered that U.S. officials refrain from issuing statements beyond mild protests. Any American response at the border, he feared, would trigger an escalation on the Communist side. Then he left his office to attend a Washington Senators baseball game. U.S. diplomats hoped West Berlin mayor Willy Brandt would not hear about Rusk's outing, nor the reaction of Foy Kohler, one of Rusk's aides: "The East Germans have done us a favor."

More than ever, East Berlin was an armed camp, CBS correspondent Daniel Schorr reported that day. Troops were needed, he added, to hold back a "sullen population." That night, Edward R.

Murrow, the legendary newsman who had left CBS to direct the administration's U.S. Information Agency (USIA), cabled his friend Jack Kennedy from Berlin, comparing Ulbricht's move to Hitler's marching into the Rhineland. He warned JFK that if he didn't show resolve he might face a crisis of confidence both in West Germany and around the globe.

Residents in the East had adapted to the arbitrary division of their city, but the character of that cleaving had changed for the worse that morning of August 13. Tens of thousands suddenly lost their jobs in the West or a chance to complete their studies, as well as freedom to visit friends, family, and lovers. Finishing their routes in East Berlin, the U-Bahn subway and S-Bahn elevated trains now discharged passengers at the border.

On August 14, Kennedy nevertheless told aides that "It's not a very nice solution, but a wall is a hell of a lot better than a war." In the same discussion, he said, "This is the end of the Berlin crisis. The other side panicked—not we. We're going to do nothing now because there is no alternative except war. It's all over, they're not going to overrun Berlin." American intelligence was almost sanguine. Kennedy's CIA briefing on August 14 dryly referred to new "travel limitations" and "restrictions" in Berlin. The next day the CIA claimed that the East German and East Berlin populations were "generally reacting with caution," with only "scattered expressions of open criticism and a few instances of anti-regime incidents." The agency may not have known that at least ten East German border guards had already fled to the West.

The administration's high-level Berlin Task Force, meeting in Washington, focused more on public relations than on countering the Soviet move with sanctions. Secretary of State Rusk stated that while the border closing was a serious matter, "in realistic terms it would make a Berlin settlement easier. Our immediate problem is the sense of outrage in Berlin and Germany which carries with it a feeling that we should do more than merely protest." Attorney General Robert Kennedy called for a boost in anti-Soviet propaganda.

On August 16, the front page of the popular West German

newspaper *Bild Zeitung* screamed, "The West Does Nothing!" President Kennedy, it complained, "stays silent." Mayor Willy Brandt cabled a forceful message to Kennedy. He criticized the "inactivity and pure defensiveness" of the Allies, which could lead to a collapse of morale in West Berlin while promoting "an exaggerated self-confidence in the East Berlin regime." If nothing was done, the next step was for the Communists to turn West Berlin into an isolated "ghetto" from which many of its citizens would flee. Kennedy must reject Soviet blackmail. At a giant rally in Berlin that evening, Brandt cried, "Berlin expects more than words! Berlin expects political action!"

Kennedy was unmoved, partly because he thought Brandt's anger was motivated as much by electoral politics as anything else. He privately referred to Brandt as "that bastard from Berlin."

WITHIN days of the erection of the concrete and barbed wire, East Germans were jumping out windows of buildings adjacent to the border along several blocks of Bernauer Strasse in the Mitte (or "middle") district, landing on the sidewalk in West Berlin. This was only possible in sections of the city where the façades of buildings marked the border. In some cases West Berlin firemen caught jumpers with their nets. A little more than a week after August 13 the first East Berliner died attempting to flee. This was Ida Siekmann, fifty-eight, who literally took flight after throwing a mattress and other belongings out the window of her third-floor apartment on Bernauer Strasse. Siekmann missed landing on the mattress and died on the way to the hospital. West Berliners were outraged. East Berlin workers bricked up windows facing West as quickly as possible.

Two days after Siekmann's fatal leap, a twenty-five-year-old tailor named Günter Litfin was shot and killed at Berlin's Humboldt Harbor. Litfin, one of the thousands of East Berliners who could no longer commute to a job in the West, had nearly finished his desperate swim to the opposite shore when he was shot in the back of the head by a border guard. Within hours, hundreds of

West Berliners gathered there and screamed their protest. Police arrested Litfin's brother and ransacked his mother's apartment. East German media launched a smear campaign against the dead man, labeling him a homosexual whose nickname was "Doll." Each guard who fired at Litfin received a medal, a wristwatch, and a cash bonus.

A West Berlin newspaper declared: "Ulbricht's human hunters have become murderers." A few days after Litfin's death, another young East Berliner was shot dead in the Treptow Canal. Within days, three more died after climbing out of windows or falling off roofs at Bernauer Strasse. In October, two more young men were shot and killed in the River Spree. Early in the Wall era, most West Berliners believed that however callous the system in the East might be, soldiers or border guards would not shoot their fellow Germans. This hope was already proving false, over and over.

Determined escapees remained undaunted. One couple swam across the Spree to the other side—pushing a tub with their three-year-old daughter in front of them.

By mid-October an eight-foot wall had replaced the barbed wire in more sections of the city. A Berlin sculptor described the Wall's disjointed, slipshod construction as appearing to have "been thrown together by a band of backward apprentice stonemasons, when drunk." Where dissidents found they could scale or blast through the concrete, GDR workers made the barrier even higher and thicker, and guard towers sprouted like mushrooms. On the side of the Wall facing west, graffiti appeared: *KZ*, the Nazi initials for concentration camp. Hundreds still made it to the West—through the sewers, in vehicles that smashed through bricks, in a train that refused to stop at the border. The Wall was both too much, and not quite enough.

Many West German and U.S. officials continued to pay lip service to denouncing the Wall while privately accepting, even welcoming it. They saw the Wall as more solution than problem, however tragic the cost for ordinary East Berliners. The main fear in the West had been that the Soviet Army—its forces far outstripping the combined Western contingent—would invade West Berlin. Only 6500 American troops occupied the isolated city compared to more than a quarter million U.S. forces in West Germany. The rise of the Wall seemed to signal that the Soviets had abandoned any plans to seize the city, satisfied for now with merely solidifying their hold on East Germany. Citizens in West Berlin, meanwhile, remained on edge. Soviet machine guns had been spotted on top of the Brandenburg Gate. Was invasion from the East really so unlikely, as the Americans now claimed? The phrase often heard on the streets was "We are sold, but not yet delivered." Residents mapped plans for the future, then often added, "If we are still here."

BY October 1961, Harry Seidel was so angry at the Communist system that he was ready to risk his life rescuing not just family members but also strangers. Options were narrowing, however. Well-designed, if filthy, escapes via the city's manholes and sewer system, known as "canalization" or "Route 4711" (the name of a popular cologne), had been discovered by the police and halted. Other schemes, requiring faked passports or forged identity papers, now drew closer scrutiny by East German border agents. Dozens of *fluchthelfer,* or escape helpers, had been arrested crossing to the East to distribute bogus IDs.

Harry Seidel, however, had studied the Kiefholz area and knew he could exploit the border's weaknesses there. Over the next three months—even as he continued competing in cycling races on Sundays—he ushered at least twenty friends (and friends of friends) over, under, or through the wires. One day he spotted a young man shouting and waving across the border to a girlfriend in the East. The man was crying. They were to be married that month, but now she was trapped. Harry promised to get her out. He did. Then he served as best man at the wedding.

His escape route was finally discovered when, as he helped a mother and baby across, the infant started crying. Guards fired on them and missed, but from then on there would be many

more GDR eyes on the Kiefholz landscape. A third line of wire
and some wooden fencing were installed. So Seidel set his sights
lower.

According to persistent rumor, the Nazis had dug a tunnel
under the Reichstag to set the 1933 fire that helped cement their
rule. Harry explored the ruins but found nothing. In the process, however, he spotted a low wall near the Brandenburg Gate
where one could leap to the East. He tried it—and froze in the
sudden glare of searchlights. Held for questioning at a nearby
police station, Harry claimed that he was only trying to escape
American soldiers, but this didn't wash with his interrogators.
When they left the room he vaulted out a window twenty feet to
the ground, found his familiar crossing at Kiefholz Strasse, and
slid back to the West, none the worse for wear.

Other setbacks were more sobering. One of Harry's fellow
escape artists, a chemistry student at the Technical University,
had offered to help another student get her mother out. As he was
cutting three wires in the dark on the outskirts of the Spandau
district, shots rang out, and the young man collapsed fifteen feet
inside East Berlin. British and West Berlin police soon arrived,
but East German border guards pointed guns at them, preventing any attempt to reach the victim. On that cold, December
day, he bled to death before the guards dragged him away.

Even far from the Wall, danger lurked. Seventeen years into
Communist rule, East Germany hosted more informants per
capita than any nation in history. Tens of thousands were aiding the Ministry for State Security—MfS for short, or "Stasi"
in the vernacular—in one form or another, paid or unpaid; compared to this, the reach of Hitler's Gestapo was tenuous. Agents
were assigned to every major factory, hospital, school, newspaper, and apartment complex, with reports forwarded to one of
the blocky buildings that made up the massive Stasi headquarters on Rusche Strasse in East Berlin. Seidel was certain that his
name now held a prominent place in those files. He also knew
that the informant network operated on both sides of the Wall.
Stasi agents were known to kidnap expatriates in West Berlin
and smuggle them back to the East.

At midwinter, while delivering newspapers from his automobile, Seidel met Fritz Wagner, a rotund, boisterous West Berlin butcher in his midthirties. They became friends and Wagner, who had a wife and two children, invited Harry to his home in southern Berlin. He knew that Harry's athletic prowess made him a natural for the hard work he had in mind: tunneling under the Wall. Wagner wished to liberate a few friends and family members from the East. He also figured to make a few bucks charging others for safe passage. Wagner, who drove a big Mercedes and maintained a newsstand on the side, was not one to miss a business opportunity. He was intrigued by Seidel's complex charisma, finding him both shrewd and devil-may-care, laconic yet temperamental, fierce but tenderhearted. Seidel, for his part, was ready for a new challenge, whatever the dangers. He embraced the tunneling concept, since other options for escape had become too risky, even by his standards. By partnering with Wagner, Harry could provide the muscle without being distracted by organizational or financial details.

Another tunnel team had just shown what was possible. Erwin Becker, a chauffeur for East German parliament members, and his two brothers had dug a shaft through sandy soil from the basement of their family's home in a remote section of East Berlin, under the Wall, coming up in a park fewer than one hundred feet to the West. Their mother warned of police activity by blinking a light in the tunnel via a switch in her home. Digging took just nine days, and late one January night ten men and eighteen women made it through the passage. *Bild Zeitung* carried photographs of the tunnel with the grinning Beckers reenacting their emergence from it. Establishing what would become common practice, the newspaper paid the Beckers for these exclusive images. The headline: "Mass Exodus from Ulbricht Concentration Camp!"

Controversy ensued. The American wire service UPI transmitted a story that included the location of the Becker home. When the West German press association protested, UPI retracted the article. This led to general agreement among West Berlin–based media that they would keep hidden key details about tunnel operations, such as precise location, the number of escapees, names of organizers, and whether or not police assisted in any way.

Seidel and Wagner took heart from the Beckers' success, although they planned to excavate in the opposite direction, from West to East. (This ran counter to nearly every example of "escape tunnels" in world history—which almost always point diggers in the direction of freedom, not from it.) For their first project they targeted Heidelberger Strasse, a narrow street in the outlying Treptow district. A high concrete barrier ran straight down the center of the street, dividing longtime friends and neighbors physically and politically. Heidelberger was now known as the "Street of Tears."

It was a scene out of a dystopian horror movie but in many ways an ideal site for a tunnel, with fewer than eighty feet separating an entry point in a basement in the West and the target in a cellar in the East. Wagner, who could barely waddle through a tunnel let alone swing a shovel in it, took a supervisory role, purchasing tools, paying off the owners of buildings on both sides of the street for the use of their basements, and recruiting a handful of workers. The digging commenced under Harry Seidel's direction. However, as a novice tunnel builder, Harry could not be certain that his aim at the cellar across the street would be true. And, as a wanted fugitive, he knew that police, border guards, or Stasi agents might very well be waiting for him when he broke through in the East.

TUNNELERS and other escape helpers were now drawing high-profile attention, even across the Atlantic. When U.S. attorney general Robert Kennedy made his first visit to Berlin in February 1962, to be greeted by starstruck crowds, he asked to meet refugees who had been smuggled from the East. William Graver, chief of the CIA's Berlin Operating Base, invited two such escapees to Kennedy's guest quarters on Podbielski Allee, a center of U.S. intelligence operations. It was early morning, and as the young men were escorted into Kennedy's suite they heard water running in the bathroom. A few minutes later, Kennedy finished his shower and emerged in his underwear, hastily grabbing and buttoning a shirt as the dialogue began. Afterward one of the students told Graver, "A German minister would never have been able to do that!"

The focal point of Kennedy's visit, a speech that afternoon at City Hall, was interrupted by flares exploding overhead, fired from the East. From them four red flags floated to earth. As the crowd booed, Kennedy exclaimed: "The Communists will let the balloons through, but they won't let their people come through!"

Harry Seidel knew little about the Kennedy visit. He was too busy, day and night, in his tunnel. The underground passage might have been relatively short, but that did not mean it was easy work. Seidel and a half dozen comrades spent several days digging down and then eastward in the cold and damp, the only illumination a few flashlights and some dangling lamps. The dirt was sandy and light but the water table was high, so the soil was wet. The young men shoveled it into large tin bowls provided by Fritz Wagner, known not always affectionately as *Der Dicke*—the Fat Man. The pots, made for transporting animal flesh in stockyards, were passed by hand or pulled by ropes back to the West, the dirt deposited in a corner of the basement.

Dig, dump, and repeat. It was like grave digging, except you had to excavate horizontally, for days, and survive in the musty chill long after light and air began to disappear. Soon one couldn't toil for more than an hour before feeling faint from lack of oxygen. Some of the men suffered from fever and hacking coughs. Seidel, strong and fit from cycling, sometimes worked for twelve hours straight, displaying few ill effects. Before long the diggers were in, or rather under, the East, as border guards armed with Kalashnikov rifles and "VoPos" (*Volkspolizei*, or "people's police") patrolled just a few feet above their heads on the Heidelberger sidewalk. Harry could hear their footsteps, some muffled chatter, a whistle. He recognized the danger, but did not respect it.

In late March, when Seidel reached the wall of the basement

in East Berlin, he created a small hole with a screwdriver and took a long look through it. Harry carried a pistol and also a fire extinguisher, ready to fill the basement with foam and drive back gunmen if he needed to make his escape. Happily, as he enlarged the hole, he confirmed that the coast was clear. Couriers from the West now spread the word to several dozen East Germans: Your tunnel is open.

The first three days of escapes were exhilarating. Harry and his helpers ushered dozens of refugees through the narrow tunnel. Friends of Wagner and the diggers made the trip for free. Those not so closely connected were referred to as "passengers" and paid a small fee directly to Wagner, who kept this policy (and the money) largely to himself. He had expenses related to the tunnel to take care of, and if he made a few bucks in the bargain that was okay, too.

The smaller you were, the easier the journey through the underworld. With a ceiling nearly three feet high, some refugees managed a stooping shuffle and made it in five minutes, while others had to crawl on hands and knees. Many were young men and reasonably fit. When they reached the brick cellar across the border at 35 Heidelberger Strasse they were muddy and exhausted, but it hardly mattered. They were in the West. To celebrate, the diggers offered them iconic bottles of Coca-Cola, a taste of what they had always imagined meant "freedom."

Then, in the final days of March 1962, Harry pushed his luck too far.

It was Seidel's misfortune that a man who lived just upstairs from the tunnel target at 75 Heidelberger Strasse happened to be a Stasi informer. His name was Horst Brieger, code-named "Naumann." Seidel had chatted with a tenant at 75 Heidelberger, and when this stranger said he wanted to flee to the West, Harry replied, *Lucky you, we've just opened a tunnel right downstairs!* Did this man know who kept the keys to the front door of the building (this might come in handy for access later)? Yes, and he directed Harry to the first-floor tenant, Brieger—who was to inform the Stasi about the tunnel the following day. Brieger even IDed his visitor as the celebrated cyclist Harry Seidel.

Now the Stasi laid a trap, which they referred to in their files as an "operative plan to liquidate the tunnel." They would allow a handful of escapees to pass unhindered, including several children with parents, while they waited for their main prey, Seidel ("organizer of the trafficking operation"), to pop out of the tunnel in the East. The chief Stasi officer at the scene ordered his comrades "to sharpen their knives for this one," meaning "they were to have their guns ready and if necessary use them."

By then, Harry could have declared victory and walked away from this tunnel. All of the refugees on Fritz Wagner's original list had made their passage to the West. But Seidel hoped to extract his wife's mother, her two brothers, and a few stragglers. Aiding him was Heinz Jercha, one of his strongest diggers. Jercha had met Fritz Wagner while working at a butcher shop in the East. After fleeing to the West, he eagerly joined *Dicke*'s tunnel project. Like Harry, the twenty-seven-year-old Jercha had a wife and young child—and an unhealthy case of idealism, reflected in his unusually bright eyes.

On the evening of March 27, Seidel followed the Stasi's script with one twist: Jercha entered the tunnel ahead of him and took the lead into the East. "You always go upstairs first, let me go once," Heinz had pleaded. After helping an elderly couple into the shaft, Jercha awaited the arrival of several students. Entering the building's first-floor corridor, he knocked on Brieger's door to ask for the keys. When the door opened he came face-to-face with Stasi commandos.

One of them ordered Jercha to surrender, shouting "hands in the air!" Jercha shone a flashlight in his eyes and ran for the basement. Seven shots rang out. Seidel, who had been waiting on the stairs to the cellar, flung open the door. Jercha staggered down the steps, through the three-foot hole in the brick, and into the tunnel. Seidel barred the door, but bullets ripped through it. Jercha, wounded in the chest, scrambled along the dark passage with Harry right behind him. Bleeding heavily, Jercha began to slow. His breath rattled as Seidel shoved him the rest of the way to the West. Pulled out of the tunnel by other diggers, Jercha cried weakly, "Help me! I am bleeding to death!" He died on the way to the hospital.

Another young *fluchthelfer*, Burkhart Veigel, who hoped to bring refugees through the tunnel the following day, arrived outside its entrance to go over details with Harry. The cyclist, who had just been grilled by West German police, seemed pale and agitated. "The pigs shot Heinz!" Harry exclaimed. "They could have just as easily hit me. This one time I let him go first—and they get him immediately!" Seidel added, "The police think I shot him. If they ask you, say that I was unarmed." Three hours later Seidel thought better of this, and turned over two pistols to the police, including a semiautomatic. He should have been arrested for possessing firearms—which were banned in the West—but cops usually looked the other way in the case of escape helpers, and did so again.

"East German guards shot and killed a West Berliner tonight," the *New York Times* reported. "Details of the shooting were not clear but police said they assumed the victim had tried to help East Germans across the Wall." East Berlin newspapers, meanwhile, hailed this defense of the motherland against "terrorists." They claimed Jercha had been shot by Seidel. Harry's mother-in-law and her two sons were swiftly rearrested. A Universal newsreel shown in U.S. theaters covered the killing under the title "Door to Freedom." Film clips showed the entrance to the tunnel, "just a stone's throw from the Wall of Hate," the narrator intoned. The "hero" Jercha had died but "his memory will live in those whose freedom he bought—with his life."

Horst Brieger, meanwhile, confirmed neighbors' suspicions that he was an informer when he suddenly started driving a new Skoda automobile. Seidel, the actual target for arrest or extermination, remained haunted by his colleague's death. Friends urged him to quit this crazy business. One told him that if he continued he should at least "never be first to break through at the other side."

Harry replied, "But that's my job." And he still had to get his mother out, come what may.

Seidel was soon to have considerably more company under the streets of Berlin. *Fluchthelfer* were planning digs at widely separated sites across the city, largely unknown to Harry but soon to gain attention from both the Stasi and the watchful West. Among the new escape artists were three students who planned to excavate far to the north of Seidel's Kiefholz and Heidelberger haunts. The months ahead would test their stamina and dedication, the ability of police in the East to expose them, and fears in the West—at the height of the Cold War—that they would succeed only in sparking a superpower confrontation. Much to their surprise, two American television networks would become deeply involved in the tunnels they would build, including one daring project that, within months, was to bring Seidel and the students—along with an intrepid Stasi operative—together on, or rather under, common ground.

2

Two Italians and a German

MARCH–APRIL 1962

News of Harry Seidel's tunnel project rocked the insular *flucht-helfer* community, provoking both fear and hope. The shooting death of Heinz Jercha was chilling, to be sure, but it had come after several nights of incident-free escapes, with dozens rescued. Since the Wall rose there had been only two wholly successful tunnels, each with a risky entrance or exit above ground. The Heidelberger tunnel showed it could be done basement to basement.

Among those taking heart were two students from Italy rooming together in a dorm at the Technical University, or TU, a hotbed of escape activities not far from the Berlin Zoo. They were Luigi "Gigi" Spina and Domenico "Mimmo" Sesta. Physically they made an unlikely pair—Spina tall and dark with a bit of a belly; Sesta short, fair, and muscular. The two had known each other since high school in Gorizia, not far from Venice, and shared wide-ranging interests: philosophy, literature, politics, economics. Gigi, after completing military service in Italy, had enrolled in an arts college, Hochschule der Künste, in West Berlin and urged Mimmo to try the engineering program at TU next door.

Reunited in Berlin, they made friends with a twenty-four-year-old arts student named Peter Schmidt, who had grown up partly in Italy and spoke Italian. He lived in the East with his wife, Eveline, and their new baby. When the border was sealed on August 13, 1961, Peter could no longer commute to the West. One week later, Gigi and Mimmo visited Peter (entering the East courtesy of their Italian passports) and urged him to consider fleeing with his family while the new barrier was still somewhat porous. Peter declined. He thought the Wall would never last—East Berliners were so against it.

Of the two Italians, Mimmo Sesta was closer to Peter Schmidt. Both were orphans. Now the two young men developed a profound bond. In the months after the Wall rose, Mimmo often visited Peter at his modest country home on the outskirts of East Berlin, talking up the need for an escape plan. He would play with the baby while Peter strummed his guitar.

It still seemed to Peter and Eveline that they would be able to build a comfortable enough life for themselves in East Berlin, in spite of the scarcities and hardship. Peter was a freelance graphic artist. Eveline liked her job in the library at Humboldt University. Their wooden cottage had an outdoor toilet, but they felt fortunate to have a house at all, as well as jobs involving little political pressure, as they waited for the inevitable dismantling of the Wall. But as the barbed wire tangle morphed into a concrete barrier in more and more places in the autumn of 1961, the sense of imprisonment grew oppressive. At Christmas that year, Peter said, "I can't take it anymore!" The search for a way out began in earnest.

After discussing various methods, including stealing a helicopter, Spina and Sesta decided that only a tunnel would suffice. A strong young man might squeeze under barbed wire, jump from a train, even scale the Wall—he might find an open sewer pipe, swim the Spree, or hide under the backseat of a car—but what about a woman and child? Peter's adoptive mother also wanted to escape. Unlike Harry Seidel, Peter had no desire to flee to the West ahead of his family. They would exit together or not at all. Adding to the urgency: Schmidt was scheduled to enlist in the East German Army before 1962 ended.

Despite its thin track record, tunneling was coming into vogue. Mimmo and Gigi were inspired that winter by a nervy project that hadn't achieved even partial success. A group of West Berlin students had started a tunnel under a remote section of the Wollank S-Bahn station, more sophisticated than earlier efforts in its use of tons of wood and iron for supports. Unfortunately, passing trains loosened the earth. Police spotted a small depression in a platform that exposed the tunnel. This produced wide media coverage on both sides of the Wall, but the students' steady progress to that point—nearly one hundred feet of burrowing—and knack for fund-raising suggested that success elsewhere was plausible.

Now, in March, and united in purpose, Sesta and Spina set out to find a site to launch their own tunnel—basement to basement, from West to East. Looking ahead, they knew they would need additional help. Neither was fluent in German and they expected that negotiations with police, city officials, and maybe intelligence operatives would arise; and Sesta was far from ready to handle the engineering duties. A neighbor in the dorm, a twenty-one-year-old advanced engineering student from Wittenberg named Wolfhardt Schroedter, seemed a perfect fit. They felt Schroedter could be trusted. He had fled East Germany for political reasons four years earlier—which always commanded respect in escape circles. Schroedter was friendly with an organizer of the fake passport scheme and knew that dozens of students involved in that endeavor were now looking for other ways to release friends and families from the East. They might be ready to roll up their sleeves and wield a shovel.

EVEN before the Heidelberger tunnel drama, Piers Anderton had issued a call for tips on any digging under the Wall. Anderton, NBC's Berlin correspondent, had covered all the escape methods favored in the first months after August 13, 1961, and knew that each was becoming ever more difficult to pull off. He needed to stay on top of the latest options.

Encouraging Anderton was his boss back at 30 Rockefeller Plaza in New York, Reuven Frank. He had helped create, and now produced, *The Huntley-Brinkley Report*, the top-rated evening newscast. Frank had come up with one of the most famous tag lines in the history of television, a sign-off for co-anchors Chet Huntley, who was based in New York, and David Brinkley, in Washington, D.C.: "Good night, Chet ... And good night, David." He had also picked their theme music, an excerpt from Beethoven's Ninth Symphony. Born in Montreal to Eastern European parents, Frank had attended college in Toronto, arriving at NBC in 1950 after a stint with a Newark newspaper. Within a decade he established the model for political-convention and election-night coverage, marked by quick shifts between an anchorman and correspondents. He was among the new breed of television producers who, having never worked in radio, placed a higher priority on moving images than on reading or reporting the news on camera. One of his pet quips: "That's why they call it tele*vision*."

Frank happened to be in Berlin with Brinkley on August 13, 1961. A few days later he instructed Anderton to follow the public mood in East Berlin closely, knowing that this could be the story of the decade. "Give us anything you find on refugees trying to get out under this new repression," he told Anderton. "Don't worry about getting permission. Go ahead and do it. I'll pay the bill." It was more a demand than a request—Anderton would liken it to a ukase—but he embraced the challenge.

Anderton was, in an era of less than photogenic TV correspondents, one of the most unusual looking. His black hair, swept back, was turning gray in broad strokes and only in front. He had sad eyes and thick lips, and was one of the rare network faces with a mustache (a bit curled at the ends) and beard. He looked like an aging beatnik, minus the poetry, the pot, and the bongos.

A native of San Francisco, Anderton was, at forty-three, a year older than Reuven Frank. His middle name, Barron, reflected his lineage back to Edward Barron, who made a fortune in Nevada's legendary Comstock Lode silver mine in the nineteenth century, and in other investments. After graduating from Princeton, Anderton had served in the Navy during World War II, then worked for the *San Francisco Chronicle* and *Collier's*

magazine and attended Harvard as a Nieman Fellow. At NBC he
wrote scripts for Chet Huntley specials, then became a foreign
correspondent. Frank felt he displayed an unusual combination
of versatility and competence. He was also aware of Anderton's
temper, which had once (briefly) caused him to resign over network meddling in one of his reports from Spain.

Anderton didn't suffer fools gladly. He had even challenged
President Kennedy at the White House back in January when, as
part of a delegation of NBC reporters, he was granted an off-the-
record meeting. When Kennedy criticized some of Anderton's reporting, the correspondent defended his work. Then he took JFK
to task for his first-strike nuclear policy in Europe. "Would you
really start a nuclear war over Berlin?" he asked, impertinently.
Kennedy said he would, if necessary.

There were maybe a dozen full-time English-speaking journalists in Berlin, but NBC boasted that it had the only fully
staffed bureau. For one pre-Wall program, *The S-Bahn Stops at
Freedom*, Anderton covered the flight of East Berlin professionals to the West via the elevated train line. For another he nar-
rated a report from inside a sewage tunnel through which East
Germans had escaped, evoking the ending of the classic film *The
Third Man*, which was set in Vienna. Whenever he crossed the
border at Checkpoint Charlie he found himself interrogated by
GDR guards, sometimes for hours. To mentally escape he would
remove from his wallet and read a T. S. Eliot poem, "La Figlia
che Piange," which closes with "Sometimes these cogitations
still amaze / The troubled midnight and the noon's repose."

On at least one occasion, Anderton had directly aided an escape plot himself.

Two *fluchthelfer* had shown up at the NBC office and asked
Anderton to lend them a pair of Japanese-made walkie-talkies.
Anderton obliged, but insisted on accompanying them on their
mission. This would be quite a scoop. One foggy evening, Anderton was driven to an out-of-the-way border zone divided only
by barbed wire. Across the barren "death strip," refugees were
supposedly waiting in bombed-out buildings. An escape helper
would cut a path (or "river," as he called it) through the wires,

creep to the building, and lead the refugees to the West. Anderton watched as one of the men, named Klaus, grasped a walkie-talkie and wire clippers and crawled out to the border. Klaus disappeared into the darkness but sent back scratchy updates via the NBC radio: "I'm through the wire. . . . Going down the slope. . . . Hut on the left. . . . Lying in a trench until the patrol passes." Then: nothing. The other man whispered into his device: "Klaus, speak . . . Klaus, come in. . . . We cannot hear. . . . *Klaus, speak.*" For half an hour they waited for a response but, except for periodic static, silence endured. Anderton never did learn the fate of Klaus.

By the spring of 1962, Anderton and other Berlin correspondents had heard that tunneling—the only escape method that kept both helpers and refugees out of sight—was gaining favor, but as yet no journalist had gotten in on the (muddy) ground floor. Anderton knew that Reuven Frank would love to sink budget resources into one. Now, in March, he asked a part-time NBC staffer named Abraham Ashkenasi to find out if any of his student friends knew anything about a tunnel, or plans for one.

WHEN the final March edition of *Der Spiegel* hit the newsstands it was clear the shadowy *fluchthelfer* community of West Berlin would never be the same. The cover line read *Flucht Durch Die Mauer* ("Escape Through the Wall") against a black-and-white image of a stern VoPo studying the West through strands of barbed wire. The article opened:

> In adventurous ways, partly above and partly under ground, since August 13 have fled around 5,000 East German citizens past Ulbricht's wall border to West Berlin. One in eight made it to freedom by means of a West Berlin student group that selflessly devoted itself to this. *Der Spiegel* reveals first details on the escape routes and the functioning of the western smugglers who dug tunnels after August 13, opened sewers, and forged passports in order to perforate the wall.

Students from almost every country in the West had taken part, with 146 arrested so far, including a pair of Americans.

The architect of all this was West Berlin's leading escape organization, known as the Girrmann Group, or Unternehmen Reisebüro ("Business Travel Agency"), in *Spiegel*'s wry christening. The Girrmann Group revolved around three administrators at the Free University (FU), a West Berlin institution founded in 1948 by GDR defectors. Two were law students, Detlef Girrmann and Dieter Thieme, and one a theology student, Bodo Köhler. All were in their early thirties and each had escaped the East as a political fugitive years before. Aided by, among others, an American student from Stanford, they focused on FU students trapped in the East before broadening their scope.

The group had operated mainly out of sight since its founding just days after the Wall went up. And no wonder. It was hard enough to carry off hundreds of escapes via checkpoints and sewers, in rafts, or by way of Scandinavia, without the press blowing your cover. Until now most in the media recognized this and held back what they knew about the Girrmann operations, encouraged by city officials who demanded discretion. After six months of secrecy, however, the three Girrmann organizers decided to go public. One reason: they had less to hide, since their early initiatives were now blocked by East German countermeasures. Secondly, after months of rescue operations, they had run up huge debts. *Der Spiegel* was happy to pay Girrmann, Thieme, and Köhler for information leading to the first inside story on escape work. The trio had expected a fee of 10,000 Deutsche Mark (at the time 4 DM = 1 U.S. dollar) but received only 6000 DM because the editors found their cooperation less than complete.

In a front-page story, the *New York Times* covered the *Spiegel* bombshell under the headline "Foreign Students Aided Escape of 600 East Berliners to West." It referred to "Scarlet Pimpernel raids" and a kind of "underground railroad." No names were revealed in either *Der Spiegel* or the *Times*, but it seemed that everyone in West Berlin knew how and where to contact the organizers. Their villa in the Zehlendorf district looked like a miniature castle and even had a catchy name: Haus der Zukunft ("House of the Future"). Besides providing office space, it served as a hostel for students from abroad, many of whom were then recruited as escape helpers. Admiration for these *fluchthelfer* was strong following the *Spiegel* piece, but not universal. On March 31, the rector at the Free University dismissed Detlef Girrmann as a director of the Student Union, charging that his escape work put the school in a sensitive political position. Even in the West.

ONE of those newly interested in chatting with Girrmann organizers was a young West German, who had left the East four years earlier, named Siegfried Uhse. Barely twenty-one, he was a barber by trade. He had a thin face and build, light-colored hair, and he dressed neatly. Slick from head to toe.

Uhse first visited the House of the Future just as the *Spiegel* piece appeared and managed to speak with the man in charge there, Bodo Köhler. He told Köhler that he wanted to get his mother and girlfriend out of East Berlin, a common request. The next day Uhse described the visit to an associate in detail: "I noticed that I was speaking to the right person. The manager told me they were not working at the moment because their last business blew off in February. Köhler wanted to know if I was West German and I said yes. We had a small chat about escape routes and I offered him my help if he needed it. He wrote down my name and address, as well as a description of my girlfriend and her address. He said he would contact me if there was anything new, but he also wanted me to tell him when I would get a new passport." Köhler, he added, "looks like the eternal student. He wears glasses with black rims. His hair is dark blond."

In the same conversation Uhse remarked that he had spotted a help-wanted ad for a hairdresser in the PX barbershop at McNair, a major U.S. Army base in Berlin. "I will try and get a job there," he added.

The person he told all this? His handler at the Ministry for State Security (MfS) in East Berlin. And that story about his mother and girlfriend? A total lie.

Uhse had served as a paid informer for the Stasi since the previous fall, after he was arrested trying to smuggle 112 cigarettes through the Friedrich Strasse checkpoint. An official report claimed that Uhse planned to deliver them for a weekly "homosexual and lesbian orgy." The Stasi had been tailing him, probably aware that he had been arrested and sentenced to probation across the border in Baden-Baden on suspicion of being homosexual—which was against the law even in West Germany.

They soon discovered that he had plied an East Berlin woman with cigarettes and wine from the West so that she would let him spend evenings in a room she rented to one of his male lovers. Uhse, who had once hoped to work as a librarian, was not much interested in politics. He had left East Berlin for Baden-Baden in 1958 simply to join his widowed mother, who worked as a kitchen aide at a sanatorium. Moving to West Berlin in 1960, he lived in a well-furnished apartment and spent nights at lounges and jazz clubs with names like the Dandy Club, Eden Saloon (favored by American tourists), and Big Apple, where he drank liberally and cultivated friends from a higher social class. He spent money beyond his means, often offering to pick up the check to impress others.

Detained by the Stasi in the autumn of 1961, Uhse was a prime candidate for undercover work on several levels. He probably still resented the West Germans for his arrest in Baden-Baden. Temporarily unemployed, he remained attached to a costly lifestyle. Now he faced a smuggling charge in the East. The Stasi felt that, in recounting his adventures, Uhse showed promise as a spinner of tales. After two days of detention, a tasty breakfast, and the promise of a regular stipend, he agreed to work as a low-level informer based in the West.

Like other Stasi recruits, Uhse had to submit a "letter of commitment" for the files. On September 30, 1961, the day after his arrest, he wrote by hand:

> *I, Siegfried Uhse, voluntarily consent to actively support the*
> *security forces of the GDR in their righteous fight. Furthermore,*
> *I pledge to maintain absolute silence to everyone about my*
> *cooperation with the forces of the Ministry of State Security*
> *and all related problems. I was informed that if I break this*
> *commitment I can be punished according to the current laws of*
> *the GDR. For my cooperation with the MfS, I choose the code*
> *name: "Fred."*

Uhse, listed in Stasi records as blond and 1.69 meters tall
(or a little over five feet six), immediately started monitoring
the West Berlin homosexual scene, but he was slow to crack
fluchthelfer circles. It was true that a Stasi informer had wrecked
Harry Seidel's tunnel, but that had been pure luck—he just happened to live above its entrance. Uhse would have to go hunting
for trouble. His big break came one night at a club when a man
told him that a student hangout called Berliner Wingolf was a
center for human smuggling. Uhse visited that club, where he
was referred to the House of the Future, inspiring that fateful
first meeting with Bodo Köhler.

Now, after Uhse's latest debriefing in March, his Stasi handler ordered him to grab that hair-cutting job at the U.S. base,
adding in his report: "Uhse is sure that the manager of Haus der
Zukunft is working with a bigger group in trying to get GDR
citizens out of the country. The manager would be interested in
Uhse because he has a West German passport."

THEY didn't yet have funding or supplies, but the three students—
Spina, Sesta, and Schroedter—were anxious to break ground.
First they had to settle on a site for their tunnel to begin in the
West and a target point across the border. The crucial consider-
ations: Would the entrance and exit be well hidden? How distant
were these two points? Was the soil loose and sandy (easier to
shovel but requiring ceiling support) or hard clay? How deep
was the water table?

Proceeding carefully in their dorm, the three plotters pored over the detailed Berlin maps obtained from sympathetic city workers, with each building numbered and underground pipes outlined. They inspected the area around the Brandenburg Gate and the Reichstag—the Stasi might not believe anyone would dare dig near the busiest tourist spots—and three other sites. Each had advantages and drawbacks relating to distance and security. There would have to be enough room in a basement to store tons of extracted soil, or a well-concealed courtyard in case they had to dump it outside or load it in trucks. From another municipal office they secured maps showing the varying water table in Berlin and learned that the area around Bernauer Strasse offered more room for error. But under which building to open a tunnel?

To their surprise, the trio pinpointed a breakthrough site in the East before they found a home in the West.

It happened by pure chance. One of Spina's friends met someone who knew an engineer from Bulgaria now living on Rheinsberger Strasse. This was the second street just across the Wall in the East, parallel to Bernauer. The two Italians visited the Bulgarian to say hello and managed to wrangle an invitation to his birthday party a couple of weeks later. On that day, while Spina distracted the host, Sesta lifted off a hook a key to the basement. Exploring the cellar, he found it suited their purposes. Recalling American crime movies where keys were stolen and impressions made in soap or modeling clay, he found a store nearby that sold plasticine. He pressed the key in the sample, then returned it to the Bulgarian's apartment. The ploy worked. A locksmith in the West soon produced a duplicate key.

With the target chosen, the options for an entry site in the West were narrowed to the stretch of Bernauer directly across the border. One site jumped out: a hulking five-story factory at Wolgaster Strasse, half of which had been bombed in World War II and neither restored nor leveled since. Behind it was a courtyard out of sight of both passersby and VoPos.

Entering the factory, Schroedter and Spina discovered that a small section upstairs was still being utilized to make swizzle sticks, the thin, plastic cocktail straws. They located the owner, a heavyset, middle-aged man named Müller. Schroedter did all the talking because Spina's German was limited. Could they use the first floor and basement as rehearsal space for their jazz band? "Do not tell me such stories," Müller scoffed, before granting them permission to use his building for their tunnel so long as they cleaned up afterward. "I come from Dresden," Müller told them, by way of explanation. "My small family porcelain business was seized by Communists. What you see here is the factory I had to start from scratch." He would not charge rent, and they could tap into the structure's electrical grid for free.

As Schroedter and Spina explored the space they got even more excited. There were rooms where diggers could sleep, hang up dirty clothes, or drink a beer, and large corners of the basement where dirt could be dumped. There was just one problem. The factory was set back from Bernauer and the Wall, requiring the longest dig of any they had considered. They would need to excavate at least 100 feet under the factory grounds and Bernauer before they even reached the border. Then they would have to toil under the block-wide "death strip" and another city street before finally, they hoped, breaking into that basement on Rheinsberger. The students calculated this would require digging over 400 feet—four times longer than any previous tunnel. And about three-quarters of it would be in the hostile East. They figured it would take at least two months.

While they recognized the added risk of water leaks and cave-ins at that length, they rarely talked about that. They were young and blessed with the corresponding bravado and sense of indestructibility. A tunnel seemed the only way to retrieve entire families such as the Schmidts. Avoiding VoPos and soldiers by burrowing under them like moles felt safer than trying to trick them at a checkpoint or hiding in a truck, or cutting through wire with armed guards and attack dogs nearby. Now all they needed were a few more devil-may-care diggers to join them. Ample supplies, including a large pile of lumber. A van to transport them. And a healthy chunk of funding (they had only 1500 DM, or about $375, among them). Plus, a few firearms. Because: You never know.

3

The Recruits

APRIL–MAY 1962

Piers Anderton wasn't the only American journalist keen to explore a tunnel with lights and camera. With upstart ABC barely in the picture, CBS and NBC battled obsessively for news scoops and muckraking specials, armed with fresh hires and large budgets. NBC's Reuven Frank had posted a sign in his office that read: *It's not important how you play the game, but whether you win or lose.* This fueled the fiercest fight in journalism since William Randolph Hearst took on Joseph Pulitzer in a newspaper war more than half a century earlier.

It was the golden age of television documentaries. They arrived nearly every week as prime-time specials or as installments in series such as *CBS Reports* or NBC's *White Paper*. Network executives wanted the public to forget the recent quiz show scandals, when winners were fixed by producers, provoking congressional probes. The legacy of broadcast legend Edward R. Murrow still clung to CBS, but NBC outhustled and outspent its rival whenever it could. Advertisers, once averse to sponsoring documentaries, now competed for spots on these prestigious shows.

CBS's man in Berlin, and Piers Anderton's chief rival, was Daniel Schorr. A Bronx native, the son of Jewish immigrants from an Eastern European shtetl (family name: Tchornemoretz), and a World War II veteran, he joined CBS in 1953 at the age of thirty-six. Two years later, after the death of Stalin and the beginning of a slight warming trend under Khrushchev, he opened the first CBS bureau in Moscow. In 1957, he obtained an exclusive interview with Khrushchev, but was soon at loggerheads with his hosts over censorship issues. When the Soviets refused to renew his visa in 1959, CBS sent Schorr to Bonn.

Schorr, like David Brinkley and Reuven Frank, was in Berlin on August 13, 1961, when he was awakened in the middle of the night. "Daniel, they're closing the checkpoints," his German cameraman told him. Schorr learned that none of his hotel's staffers from the East had arrived for the midnight shift, not a good sign. Rushing to Potsdamer Platz in his silver Mercedes, he found soldiers and guards unspooling barbed wire and closing off streets. As the sun climbed in the sky, Schorr recounted on camera in his distinctive baritone, "My cameraman and I were arrested and held in a police station for ninety minutes," their footage briefly confiscated. (Piers Anderton had also been detained.) Schorr then slipped the footage on a flight to Frankfurt, where it made the next Pan Am airliner for New York.

The next day, Schorr reported, "Small numbers of East Germans are continuing to break through the Communist cordon to West Berlin." These included a young engineer who kicked a cop in the stomach, another man who seized a guard's carbine, and "one who stepped on his car gas pedal and rammed through." The following day Schorr was on Bernauer Strasse to witness the first prefabricated concrete slabs being lowered into place, "as though to build a wall" (making him perhaps the first American correspondent that week to use the word "wall"). "We might have been willing to go to war to defend our right to stay in Berlin," Schorr intoned, "but can we go to war to defend the right of East Germans to get out of their own country?" That day, a photographer captured one of the decade's indelible images, published overnight around the world: an East German border guard leaping to the West over a section of low-slung barbed wire along Bernauer Strasse.

Two months later, in October, Schorr provided riveting

coverage of a frightening U.S.–Soviet confrontation at Checkpoint Charlie, the main American-run crossing point at the Wall. The chief of the U.S. Mission in Berlin, Allen Lightner, had refused to show his papers to GDR guards while on his way to the opera in East Berlin; as a top diplomat he was supposed to deal only with the Soviets and feared setting a precedent. Within hours an absurd superpower shooting war seemed ready to break out as tanks faced off across the border. Schorr captured the tension as "American and Russian fighting men stood arrayed against each other for the first time in history." But he also highlighted the "weird" scenes of West Germans bringing U.S. soldiers flowers and snacks under the glare of Soviet searchlights from across the border. "What a picture for the history books," he predicted.

Schorr had come to believe that World War III might very well start in Berlin—you could feel and even hear it, as Soviet MIGs flew low to create a sonic boom just to keep residents on edge. Kennedy was not looking for a war, Schorr believed, but it was hard for any U.S. president to risk losing face in stepping away from a crisis. The outlook for West Berlin looked grim. It was, Schorr felt, locked in the arms of East Germany and something had to give.

Like Piers Anderton, Schorr had been ordered by his bosses to find out what life was like in East Germany, but the irascible newsman needed little encouraging. Despite his success, Schorr maintained something of an inferiority complex, born of growing up poor, overweight, and fatherless. A self-professed outsider, he was, he admitted, "pushy," not always liked by his superiors at CBS and prone to courting controversy, but he was also dogged in pursuing a scoop. Late in 1961 he began working on a documentary about the East, *The Land Beyond the Wall*, and somehow got permission to spend more than two weeks in the city of Rostock (with an East German chaperone, of course). The *New York Times* called it "a journalistic coup."

Schorr's visit was nearly finished when he received a shocking offer: Communist leader Walter Ulbricht would sit down with him in Berlin for his first-ever filmed interview with an

American. That day, as Ulbricht provided vague, long-winded answers, Schorr interrupted him with follow-ups, upsetting a man unaccustomed to impertinence. Ulbricht finally stood up, pointed a finger at Schorr, loudly accused him of "provocations," banged on a table, and stalked out. Schorr thought, *What a great ending*. When West German television aired the footage, a local newspaper carried an image from it on its front page under the headline "America Laughs at Ulbricht."

Now Schorr set his sights on another coup: filming a major tunnel operation, perhaps even crawling through the passage with a cameraman to the Communist side. Like Piers Anderton, he put out word through his contacts that CBS was ready and willing to cover such an enterprise. Schorr knew that filming a tunnel escape could put him in danger of physical harm, and he faced an additional risk: opposition from the White House. Schorr's boss, CBS news director Blair Clark, was a Harvard classmate of the President and remained friendly with him, maybe a little too friendly. Clark had told Schorr that the White House was continually unhappy with the CBS correspondent's reliance on leaks from West German officials that painted U.S. actions—or inactions—in a poor light. Clark revealed that at a recent White House dinner, JFK had leaned over and advised, "Blair, that Dan Schorr in Germany is a pain in the ass—why don't you pull him out of there?"

AMERICAN TV correspondents were still searching for their first tunnel to cover from the inside, but Hollywood already had one. MGM announced that it would soon start filming a major movie in Berlin based on the true story of the year's only completely successful tunnel: the Becker opus, from back in January, which had freed twenty-eight East Germans.

Walter Wood, who had just produced the well-regarded *The Hoodlum Priest*, would perform the same duties on this film, and had hired Don Murray, Christine Kaufmann, and Werner Klemperer for the leading roles. Kaufmann, just seventeen, was having a rather public romance with actor Tony Curtis, married at the time to actress Janet Leigh of *Psycho* fame, guaranteeing extra publicity around the film. A director had not yet been chosen. Filming would commence at the famed UFA studios in Berlin's Tempelhof district, where classics by Fritz Lang and Josef von Sternberg, as well as notorious Nazi propaganda films, had been created. Mayor Willy Brandt gave Wood permission to shoot at the Marienfelde refugee camp.

The budget was set at a substantial $500,000, with a thirty-five-day shooting schedule later that spring. A fake tunnel would be built on the studio lot, but cameramen would also shoot on the street to capture real-life Wall atmosphere. No doubt the VoPos would serve as reluctant, unpaid extras. One of the tunneling brothers, Erwin Becker, had been hired as a consultant. Hedda Hopper, the Hollywood gossip columnist, claimed that Becker "was found in a transit camp by Wood, is now under guard, and telling his story for the screen." MGM feared that Becker would be kidnapped by the Stasi and returned to East Berlin before the film could be completed. The title of the movie: *Tunnel 28*.

WHILE Hollywood planned its fictionalized version, the longest and most dangerous real-life tunnel project was almost ready to break ground. The only German in the three-man tunneling operation, and its youngest member, Wolf Schroedter, had quickly proved indispensable. He stood out even at a glance: Tall, thin, with a blond crew cut, he was easily distinguished from dark, husky Gigi Spina and compact Mimmo Sesta. Through the Student Union he began to identify potential recruits. Schroedter also took charge of finding a van, knowing he would serve as chief driver (he was the only one on his team with a license). If that wasn't enough, he knew that he would likely do much of the negotiating with government officials who sometimes donated small sums to *fluchthelfer* under the table. And as the only organizer who owned a gun, Schroedter was, you might say, more or less in charge of security as well.

Thanks to a tip from Egon Bahr, a top aide to Mayor Brandt, the tunnelers obtained 2000 DM in seed money from one of the German political parties. Their big funding break, however, came via the Italians. The adoptive mother of their friend Peter Schmidt told them that she had a tidy sum of cash in a West Berlin bank, at least 3000 DM, courtesy of her late husband. She offered to give Spina or Sesta power of attorney to secure the funds. There was, however, a problem: Keeping money in a Western account was now illegal for GDR citizens. If one of the Italians was caught at the checkpoint with a signed document for the bank, she might be arrested.

Undaunted, the Italians devised a scheme: They took one filtered cigarette out of a full pack and removed the tobacco. Schmidt's mother signed over power of attorney on a small sheet of very thin paper, which they had cut and dyed to match the cigarettes. After returning the loose tobacco, they rolled the cylinder, inserted it in the back of the pack, and set off for the West. If they were hassled at the border they might be able to smoke the evidence out of existence.

The trick worked. With this infusion of cash, they bought a secondhand Volkswagen van which, lacking windows on the back or sides, would be ideal for hiding tools, supplies, and co-workers.

Schroedter, meanwhile, had secured two more diggers. Joachim Rudolph and Manfred Krebs lived right across the hall in the TU dorm. The pair had been friends since childhood. Schroedter felt they could be trusted because they, like him, had fled the East. Rudolph, formerly an engineering student in Dresden, had spent many sleepless nights in deciding to leave the GDR the previous September. Besides disrupting his studies, it meant leaving his mother, with whom he had survived a previous escape—from the Red Army invading Germany in 1945. He and a friend ultimately crossed the border in the remote section of Luebars in northern Berlin, taking four hours to crawl eight hundred feet over the earth, across a small river, below guard towers.

He and Krebs already had a notable friend in the *fluchthelfer* movement, though they didn't yet realize it: Harry Seidel. The three were former schoolmates. On a bike tour of the Harz Mountains in 1953, Rudolph had marveled that Harry, who was fourteen at the time, was able to generate such power with relatively thin legs. Four or five years later, Harry joined a top cycling club and became famous. Rudolph would attend the races and cheer. Seidel pedaled like a madman and took crazy risks. He would crash into a side wall, leaving Rudolph convinced, *It's over for him*. Then Harry would reappear and finish the race.

Rudolph had since lost track of his old friend. He did not know that Harry had led the Heidelberger tunnel team. But in entering the subterranean escape world he was now following Seidel into a different sort of high-stakes arena.

AS months passed, more and more East Berliners abandoned hope that the Wall would soon crumble. A woman who lived on a high floor near the border complained in a diary entry that the Wall at Bernauer had become "the number one tourist attraction" in Berlin. From a window she could watch the tourist buses roll by. "Oh, how gladly we would just be ignored," she added. "What a ghastly time. Our lives have lost their spirit." The authorities "will do with us what they like," she wrote. "Bow your heads, friends, we are all become sheep."

Although the seeming permanence of the Wall discouraged many, it inspired others to act. Berlin police and newspapers in the West monitored episodes on both sides of the border, and in April they covered the usual broad spectrum:

- Philip Held, nineteen, a trained electrician, drowned in the Spree, attempting to flee. His mother was not notified for two weeks; by that time his body had been cremated.

- A nine-year-old East Berlin boy ran away from home and jumped from the roof of a five-story apartment house on the border into a fire brigade's net on the West Berlin side. The boy was rushed to a nearby hospital for treatment of possible back injuries. He told police he was about to be taken from his mother, who was jobless, and sent to an institution. He wanted to live with relatives in West Berlin instead. A second boy on the roof was dragged away by a guard before he could jump. Both boys had school knapsacks strapped to their backs.

- Three young East Berliners tried a daring escape at the Heinrich-Heine border crossing. Klaus Brueske added concrete and gravel to his truck to give it added heft. Just after midnight, having fortified himself with a few drinks (a common practice for escapees), he piloted the vehicle straight through two barriers and barely into the West, but was shot by border guards en route. The truck crashed into a barrier. As Brueske sat wounded, the gravel poured into the front seat, slowly suffocating him. His two companions survived. A West German newspaper the next day ran the headline "He Died Driving Them to Freedom."

- Horst Frank, a horticultural worker, was killed at the border in the Pankow district. Late at night he and a friend had ducked under barbed wire and crawled across the death strip, dodging trip wires and guards for four hours. They had reached the last wire when Frank was shot by three guards. His friend made it to the West.

Even after dozens of shootings, Western analysts struggled to find any pattern revealing the precise "firing orders" for East German guards. After interrogating several former guards who had defected, the U.S. Army came up with what it thought was a list, including: "Defectors will not be permitted to cross the border alive. . . . No punishment will ensue for firing in the direction of West Berlin if a defector is hit or if West Berliners are attempting to cut the border fence. . . . No firing at children, pregnant women or old people is permitted. . . . Tear gas may be used, but it is not to be thrown into West Berlin."

After any fatal shooting at the Wall, known officially in the

GDR as a "corpse case," the Stasi took command. The dead were hauled to state-run medical facilities for autopsies, the results subject to falsification as needed. The state's main goal was to keep episodes hidden from East German citizens and West German media. When Stasi operatives visited family members of the deceased, they would, whenever possible, simply say that their loved one was "missing." This might elicit a response revealing why the culprit attempted to flee and whether he or she had any help. When the Stasi did confirm a death the true cause would usually be hidden even from relatives—the victim had simply drowned or fallen. Sometimes they admitted there had been an "accident," with the refugee invariably blamed for creating a "border provocation." Family members would be ordered not to tell anyone. By that point the Stasi may have already delivered the remains to a crematorium and ordered the ashes buried. If a funeral service was allowed, days or weeks later, Stasi operatives always attended. Agents might keep the family under observation indefinitely. Family members might lose privileges or even their jobs.

Dozens of border guards and soldiers had joined those attempting to flee. On April 3, a guard at a checkpoint tried to reach the West, but a second guard set his dog on him and shot him twice with his pistol. A few days later, two GDR army officers took flight in the dead of night. One of them was nineteen-year-old Peter Böhme, who was unhappy about being forced to serve in the army as a penalty for juvenile infractions. A manhunt ensued. At the border near Babelsberg, he shot and killed a VoPo, then was fatally wounded by another guard. Böhme's companion made it to the West.

Outside the GDR's propaganda bubble, gunfire at the border achieved greater notoriety in Western media, inspiring at least one fledgling fiction writer. A young British intel operative, then working out of the country's mission in Hamburg as a political consul, was writing a novel that ended with his hero, a British spymaster, getting shot at the Berlin Wall while aiding an escapee. David Cornwell was penning the book under the pseudonym John le Carré.

Amid the upsurge in violent incidents, the CIA had launched a study of the chances that anger in the East would erupt into open rebellion. The classified National Intelligence Estimate that emerged cast grave doubts on this scenario. Discontent had increased since the coming of the Wall but "there is no evidence of significant organized opposition," the report observed. A major revolt, fueled by "revulsion" for Ulbricht, could develop from "local outbreaks" but "we believe that the presence of Soviet military forces and the memories of their use in past repressions will deter the people from rising up in revolt."

Morale was grim in East Berlin, to be sure. The GDR regime expected the Wall to make the populace more malleable, according to the report, but instead it "had an effect opposite." It produced severe psychological effects even beyond losing access to jobs and families, engendering a feeling of hopelessness, as demonstrated by a startling increase in the suicide rate. A depressed economy wasn't helping matters; while the standard of living in East Germany exceeded that of others in the Soviet bloc, the past two years had seen increasing demand for quality food and consumer goods—and a decrease in supply. There was little prospect of improvement in the near future. Industrial and farm production had stalled. That meant more rationing.

Still, most industrial workers maintained Germany's traditional respect for authority and work. Students reflected more virulent anti-regime sentiment, but the state "has moved speedily and ruthlessly against youthful leaders," leaving little hope for any sort of broad protest movement. With all this in mind, the report concluded that an uprising in East Germany "would not be successful unless launched in conjunction with Western military operations."

Despite the tensions along the Wall, young guards on both sides, mainly conscripts, sometimes chatted, commiserated, or tossed cigarettes over the concrete or through barbed wire. On one occasion an East German guard passed a note to his opposite, asking if he would please heave over the Wall a package of "seamless tights," then in short supply in the East, for his girlfriend. "Size 9½" and "not too bright" in color, he requested.

"Thanks in advance!" The West German guard replied with a note written on the back of a wallet calendar, asking for an address to mail the tights. Or he could "throw the stockings over when you get back here." Signed: *Ever friends!*

The East German guard wrote back: "Unfortunately, I cannot specify address. But please watch when I stand guard here again." Signed: *Your friend!*

THE Girrmann Group, ever the innovators among West Berlin's escape artists, had come up with a new racket to replace their scuttled fake passport scheme. Having deemed tunnels unreliable (to say the least), they stepped up efforts to transport refugees to the West hidden under the dashboard or backseat or in the trunk of automobiles. Anyone in the East could quietly ask a friend or family member in the West to fill out an application for this service, which included questions about the would-be refugee's profession, color of hair and eyes, where they could be reached and at what time, and who might serve as a courier. Applicants provided a password a courier could use to communicate with them. Then there was this query: *Who is suspicious?*

Meanwhile, the Girrmann Group's most suspicious character went unnoticed. Indeed, Siegfried Uhse's Stasi career was looking up. He had secured that job cutting hair at McNair, the U.S. Army base. Officials there even gave him a pass to visit the soldiers' barracks. He had also met three possibly useful sources in West Berlin bars and clubs. One of them was a German woman who knew a lot of Americans. Uhse's Stasi handler, named Lehmann—apparently an expert in the art of spycraft romance—suggested that he bring her chocolates and flowers. Lehmann also asked Uhse to secure a map of the barracks and taught him how to use a dead mailbox, where an operative could retrieve or deposit documents and spy equipment.

Until this point, the Stasi had to pull strings to get Uhse through the checkpoints so he could meet with his minders in the East. Now he was about to obtain a new West German passport, which would allow him to enter the East far more easily. Ironically it was Bodo Köhler who had pressed him to get the passport so he could work as a courier for the Girrmann Group. It was all coming together for the dandified young man. (He even bought an expensive tablecloth for his mother.) When Uhse again met with his handler, Lehmann reported:

> We talked about his potential promotion. I told him we had reached a new stage in our work, and we would start working together in a more "professional" way. We discussed what his new obligations would be. Mainly we talked about the fact that the work is voluntary and should be done honestly, but he is under obligation to keep quiet about all his contacts to the MfS and the GDR. . . . I explained to him that we would be providing for his security, but that he should not talk to friends or relatives, even close ones, or make ambiguous remarks. If he does not hold to his obligations he would be held to account according to the laws of the GDR.

Uhse agreed to the MfS demands, promising to "follow the rules strictly," Lehmann wrote. With this he was promoted, along with a hike in pay to 100 DM every week or so. (He also got to select a new code name, dropping the undignified "Fred" for the jaunty "Hardy.") Uhse vowed to "bring more qualified and interesting results, and to be more conscientious in doing his assigned tasks," in Lehmann's telling. Material gains were "not his main motivation." The young man admitted that he did not understand politics. Lehmann noted that he "has an open mind, but he needs to be corrected on some questions." The Stasi spied on everyone, not least their own informers, so with Uhse's promotion came close "observation," as Lehmann put it, to make sure he was following instructions. Uhse's new priority: to "contact the manager of the Haus der Zukunft about escape routes."

. . .

THEY had waited long enough. The three students had found an entrance and exit point, and added two diggers, Rudolph and Krebs. Now they opened the first hole in the floor of the factory basement with a pickax, marked a rectangle on the concrete, and began chipping away. They needed to dig straight down, enlarging the initial hole to a seven-foot width. Then they would angle the excavation to a depth of about thirteen feet, and as deep as twenty feet for most of the length of the tunnel, before digging upward near the end, deep inside East Berlin. The chamber heading East would be about forty inches high and wide.

Spina, with a small, 8 mm movie camera, recorded the first moments, despite the dim light, for history—and perhaps for sale later. *Der Spiegel* had paid its Girrmann sources for their recent "Business Travel Agency" cover story; perhaps editors there, or maybe a movie studio or TV network, might spring for an exclusive on this herculean tunnel project. Knowing they would need more funds, the Italians had decided to take photos and shoot amateur film throughout the project, in hopes of finding a distribution deal, with payment in advance.

It didn't take long for the diggers to realize they needed reinforcements. Excavating the vertical entry shaft was taking longer than expected, and they could never mount an around-the-clock operation with just five diggers. One addition was easy: a friend of the Italians named Orlando Casola, a quiet young man who wore sunglasses almost everywhere. But the three organizers had a difficult time finding others they could trust.

Several days later, two men who had heard rumors about a new tunnel operation approached the Italians. One of them was Hasso Herschel, a student introduced to them in the cafeteria at TU; the other, his friend Ulrich Pfeifer, was a full-time construction engineer. Herschel, a former political prisoner—survivor of the East German *gulag*—enjoyed spinning scenarios for daring tunnel escapes over a few beers, feeling like the plots were straight out of *The Count of Monte Cristo*. He made an immediate contribution. Hasso had been digging water lines at a cemetery to earn money and knew where the shovels, wheelbarrows, and tools were stored. That night he crept into the cemetery, liberated the equipment, and piled the items into the VW van. Schroedter hauled them to the factory.

The new recruits had met a few years earlier. Uli Pfeifer, like Hasso, had grown up partly in Dresden, surviving both the devastating firebombing in World War II and the deprivations afterward. Later, one of his friends courted Hasso's younger sister, Anita. On a visit to the Herschel home in 1957, Uli spotted a photo of a dark-haired young man prominently displayed on their radio console. Anita explained that this was her brother, Hasso, who had been in prison for more than three years. He was first arrested in Berlin in 1953, at the age of eighteen, for taking part in the popular uprising against the Communists' economic policies and police state rule. A national-caliber swimmer, Hasso might have been, like Harry Seidel, the sporting pride of the GDR. Instead he was held in a single jail cell with twenty-two others for several weeks with little to eat and no change of clothes. He decided right then that one day he would flee this half of his native land.

When Hasso emerged from jail, he learned he had been barred from high school. After getting his diploma via night school, he was admitted to the German College of Politics in West Berlin and rented an apartment there. Freedom! But then, while visiting his parents back in Dresden, he was arrested and charged with violating the "law for the protection of inner-German trade." He had sold in the West a camera, a typewriter, and binoculars from the East. On this charge, or pretext, he endured four years under degrading conditions in labor camps. He was released in 1958 at the age of twenty-three.

On his return to Dresden, Hasso met Uli Pfeifer. He also did some freelance work for West German intelligence and then the CIA. For the Americans, he spied on an East German base, jotting down how many men and vehicles entered and exited, even digging in the trash for letters that might reveal secrets. After about nine months of this the Americans gave him money for a plane ticket to Berlin. He enrolled in college again, but in

August 1961, like thousands of others, he was trapped in the East
by the Wall.

Pfeifer, employed as an engineer in the East, was making
plans at the same time to flee with his girlfriend. One night in
September 1961 he staggered through a sewage tunnel to the
West, crouching amid the noxious odors, with his girlfriend soon
to follow. But as she headed for the opening, the Stasi plugged
the tunnel. She was arrested a few days later and sentenced to
seven years in prison.

Imagine Uli's surprise one Sunday morning that October
when he opened the door at his mother's home in the Charlottenburg district, and found Hasso Herschel standing before him.
They fell into each other's arms. Hasso had crossed the border
the day before at Checkpoint Charlie using a fake Swiss passport
(arranged by the Girrmann Group), after cutting his hair and
donning glasses to match the passport photo. His only regret was
that he had not scaled the Wall instead, giving the Communists
the finger as he jumped to the other side. That was essential
Hasso in one image.

Now, five months later, the tunnel project at Bernauer Strasse
had gained two highly motivated diggers. Pfeifer had no hope
of retrieving his imprisoned girlfriend but was still enraged by
what had happened to her. He couldn't imagine locking up a
twenty-two-year-old woman for much of her young adult life
simply because she wanted to leave a totalitarian state and join
her boyfriend. The convivial Herschel, who was now studying
political science at the Free University, was still angry about
serving so many of *his* formative years behind bars. He had a
concrete goal, however: getting his kid sister, Anita, along with
her husband and their baby, to the West. Anita had wanted to
leave with Hasso the previous autumn but he had talked her
out of it, saying, "One of us at a time—you are the youngest, we
cannot leave mother alone yet, and it will be hard with the baby.
But I will take care of it." Now he vowed that he would not shave
again until he did so. Hasso likened himself to Fidel Castro who,
hiding in the mountains in Cuba, had vowed not to cut his beard
until his rebels took Havana.

With the new recruits lending a hand, the diggers finally reached the depth they were seeking. Someone drew a rectangle on the side of the solid clay formation facing East. A large electric drill pointing into the clay eased the way. They were on their way to the Wall, and beyond.

4
The President

MAY 1962

President Kennedy had survived a shaky economy, the Bay of Pigs disaster, and charges of not doing enough to end racial segregation to retain much of his personal popularity. The three television networks played a key role, enabling him to become the first president to speak directly to Americans on a frequent basis—live, unedited, and without a filter. His press conferences, staged twice a month on average, were televised in full by the networks, a first for any president. Then again, the television age had never experienced a president this young, handsome, and witty. Even in the afternoon the broadcasts drew millions of viewers, helping to keep Kennedy's approval rating in polls hovering around 75 percent.

Early in his first press conference this May, in the usual impressive setting—the new auditorium at the State Department—a reporter asked him how he thought the media had treated him, beyond these periodic TV specials. "Well, I am reading more and enjoying it less," he quipped, drawing laughter from the room, "but I have not complained nor do I plan to make any general complaints. I read, and I talk to myself about it." Journalists, he added, were doing "their task, as a critical branch, the Fourth Estate, and I am attempting to do mine, and we are going to live together for a period—and then go our separate ways," drawing more laughs.

Many previous presidents had a love-hate relationship with the media, none more so than John F. Kennedy. His aide and chief speechwriter Ted Sorensen was often struck by the duality. It seemed to Sorensen that his boss regarded reporters as his natural friends but the news outlets they worked for as his natural enemies—almost as if the journalists' sympathetic words were twisted against him before they were printed or broadcast. Kennedy couldn't understand why a newspaper like the *New York Times* supported his administration overall but hammered its shortcomings, month after month, in editorials. "I'm convinced," he told Sorensen one morning, "that they keep in stock a canned editorial on 'lack of leadership' and run it every few weeks with little change." To his friend Ben Bradlee, the Washington bureau chief for *Newsweek*, he complained, "When we don't have to go through you bastards we can really get our story to the American people."

While his televised press conferences promoted his popularity, Kennedy's honeymoon with much of the press had not survived his first spring in office. In April 1961 he asked media managers to keep secret the (misguided) plans for the CIA-backed Bay of Pigs invasion by Cuban exiles. Just one outlet, the *New York Times*, published a vague report, but that was enough to get JFK's blood boiling. Two weeks later, in a speech to the American Newspaper Publishers Association, he boldly asked "every publisher, every editor, and every newsman" to "reexamine his own standards, and to recognize the nature of our country's peril." The United States was threatened around the globe by the Communists and "in time of war, the government and the press have customarily joined in an effort, largely based on self-discipline, to prevent unauthorized disclosures to the enemy." At such a time courts have held that "even the privileged rights of the First Amendment must yield to the public's need for national security."

The Communist threat required an unprecedented change in outlook not just by the government but by every news outlet. Each democracy, the President said, recognizes necessary restraints on the media—and the question in America was "whether those restraints need to be more strictly observed." He railed against leaks that might tip off enemy powers. These leaks might pass the test for journalism but not for national security, and Kennedy wondered aloud whether additional tests "should not now be adopted." He urged his audience to give it "thoughtful consideration" and reexamine their "responsibilities."

When excerpts from the speech were published, media commentators—with or without thoughtful consideration— rejected what many felt were thinly veiled threats to impose new controls if the call for self-restraint was not heeded. Under a headline "The Press: No Self Censorship," *Time* magazine called the speech "ill conceived." Even some Kennedy aides, such as Arthur Schlesinger Jr., felt he had gone too far. The President backed off, but his views on press irresponsibility festered.

NEARLY every morning at the White House, Kennedy received a few concise pages from the CIA known as "The President's Intelligence Checklist." This breezy briefing offered nuggets of news and updates from hot spots around the world, from Laos and Vietnam to Cuba and the Congo. A lead item one day in May read:

> Tension along the Wall, already high as a result of shootings and explosions that have attended attempts by East Germans to escape to West Berlin during the past week, is likely to go even higher. Passions over the refugee problem are being whipped up on both sides by shrill propaganda exchanges. Mayor Brandt has authorized his men to use their guns if necessary to assist escapees, and the East Berlin security forces seem to be more trigger-happy than usual and have been heavily reinforced along the entire Wall. More clashes between the opposing police forces seem inevitable.

"The East Germans seem intent," the item concluded, "on stemming the flow of escapees which is still averaging about 50–60 a week. They may even be seeking an opportunity to respond to an escapee incident with enough force to torpedo Mayor Brandt's publicly announced policy of active assistance to escapees." On May 21, the CIA assured the President that Fidel Castro's purge of hardline Communists from his government displeased the Kremlin and "may lead to serious problems with the USSR." The agency did not know that five days earlier Khrushchev had decided to send nuclear missiles to Cuba.

Reports on Berlin often dominated the "Intelligence Checklist." Berlin was not only the political flash point of the Cold War, it was also the front line of the intelligence battle, one of the few arenas where the United States and the USSR came into direct, one-on-one conflict. It was an espionage landscape perhaps unique in history. Both of the operational centers were in one (formerly united) city, and until 1961 each side had remarkable access to the other. The period from 1945 to 1961 was considered the heyday of the Berlin intelligence war, spy vs. spy, the stuff of numerous novels and movies. The Wall significantly hampered these operations, and rather quickly Berlin came to play a less important role in superpower espionage. In 1962, all sides were still adjusting to the new reality.

U.S. and Soviet intelligence shared two basic aims in Berlin. One was straightforward: information. Each side wanted to learn as much as it could about the military capabilities, political trends, and economic conditions of its adversaries—both the German opponent and its superpower patron. The Wall seriously disrupted the intricate networks of agents and informants.

The other goal, as in spy games everywhere, was self-referential: preventing subversion and conducting counter-intelligence operations against foreign agencies. With its unprecedented number of informants covering both East and West, the Stasi was masterly in this regard. Top GDR officials were driven by the fear of rebellion, fostering the paranoia and obsession that came to characterize the MfS. East German citizens were viewed as "unreliable" and vulnerable to Western propaganda, and thus in need of tight control. The Stasi even infiltrated the GDR's military and police agencies.

Under the direction of seemingly all-powerful Erich Mielke since 1957, the Stasi had built a loathsome reputation. Its massive Hohenschönhausen prison in Berlin was much feared, even though the torture carried out there was now more psychological than physical. Still, MfS's effectiveness did not take full shape until the Wall was built, when its ability to physically control and monitor the population soared. Dissidents once fled to the West overnight, but no longer. Some in Western intelligence began to view the GDR as a "dictatorship of the Stasi," as if the MfS controlled everything. This underestimated the power of the ruling SED party. In fact, the Stasi's motto was *Schild und Schwert der Partei* ("Shield and Sword of the Party"). Arguably the MfS's strongest contribution was not the information it gathered but the aura of secrecy and omnipotence it generated on behalf of the SED, which allowed the party to crush free expression and remain hard-line Stalinist.

Perhaps the Wall's greatest impact on Berlin intelligence was the sudden stemming of the refugee tide. Before August 1961, both East and West spy services had taken advantage of the mass movement of refugees through Berlin. GDR agencies inserted operatives in the refugee flow and thereby gained easy access to the West. But there was this flip side: East Germans who made it to the West were required to register at the Marienfelde refugee camp, where intelligence officers from the United States, France, Great Britain, and West Germany were free to question them.

But now, with far fewer refugees making it to the West, Marienfelde seemed in danger of becoming a ghost town. This made U.S. intelligence all the more eager to chat with any East Germans who did make it, via tunnels or any other route. They counted on escape helpers to supply them with fresh refugees who might possess current evidence from the East. At the same time, of course, CIA officials feared that organized escape operations might provoke a U.S.–Soviet confrontation.

.   .   .

AFTER several days of digging vertically, the youthful team in the cellar of the swizzle stick factory had finally made the fateful turn east. The work was only getting harder. For one thing, it was colder and damper in the chamber than they had expected. Even on very warm days above ground the temperature in the tunnel did not get much above 55 degrees Fahrenheit. More critically, as they started the long horizontal portion of the project, their shovels met stubborn resistance. The soil at this depth was not the sandy stuff near the surface but a thicker and far heavier clay. That slowed both the digging and the dumping.

At the start they had rigged up a bucket-rope-winch pulley system that carried the dirt up to the basement floor for depositing in a corner. The clay made this a truly weighty task. Moreover, to force the spade even a few inches into the dirt, a digger had to lie on his back and push hard with his feet on the blade, then turn slowly and awkwardly to spill the contents—or what was left of it—into a small cart. The electric drill helped loosen the wall of clay at first, but the diggers knew that once they approached the border they would have to abandon that noisy tool.

After advancing a couple of dozen yards, they started placing thin sheets of wood on the dirt floor, then ran a steel rail down the center and added a single rubber wheel to the cart. The digger would fill the cart, not much bigger than an apple crate, with about twenty-five pounds of dirt, then shout or pull on the rope to signal a man at the mouth of the tunnel to haul the pile to him using a crank. When the load arrived, he would drop it in a wheelbarrow, and return the cart to the front. The hands of diggers soon turned sore and blistery. Over and over they repeated this dig-and-dump routine, lit by a string of lights installed by Joachim Rudolph and plugged into the factory's electrical grid. Finally they obtained hand-cranked Wehrmacht field telephones (vintage World War II) to communicate from front to back in the tunnel.

One good thing: the firm clay meant the sides and roof of the tunnel were more likely to hold. Still, the diggers decided to construct strong supports. With some of their dwindling cash they purchased a small pile of lumber (unloaded in the factory's courtyard, out of sight of the East) and started sawing it into logs about a yard long and four inches in diameter. Uli Pfeifer had suggested a triangular shape for these supports. Every few feet they would install two logs meeting at the peak of the tunnel. Boards would then be fitted along the sides. Even so, the possibility of a cave-in or sudden water leak—ancient pipes ran overhead—threatened their safety.

Daily progress was measured in feet, not yards. Advancing six feet marked a good day. The wood was already running low, and soon they would have to install some kind of pipe to carry air to the far end of the cavern. (Mimmo Sesta kept testing the oxygen levels by lighting matches.) Clearly they needed to turn more hands to labor, but it wasn't easy finding volunteers for this dangerous, backbreaking work. Another obstacle: nearly everyone in or around this wing of the *fluchthelfer* community was some kind of student. Dedicated digging might require abandoning their studies for a semester. The other wing of the escape community—represented by Fritz Wagner, Harry Seidel, and a pair of brothers named Franzke—were blue-collar types. The two camps, students and workers, rarely interacted, limiting the pool of fresh recruits.

Fortunately, several other young adventurers were looking to get their hands dirty.

First up: Joachim Neumann, yet another engineering student at the Technical University. Near the end of 1961 he had fled the East using a fake passport. In April he arranged his sister's passage to the West, hidden in an automobile. Now he wanted his girlfriend, Christa, to join them. Another student, known as Oskar, who had worked on the failed Wollank S-Bahn tunnel, knew colleagues from that dig who would join any new project. One day in May the pair visited an official at the state Ministry for All-German Affairs known to be sympathetic to escape helpers. He said he couldn't fund their proposed tunnel, but he knew of one that had already broken ground—and he'd heard the organizers needed help.

A few days later, the two students returned to that office and were told they could meet the tunnel organizers one evening at a restaurant in the Wedding district. How to ID the tunnelers? One of them sported a beard—a rarity in Berlin in those days. Sure enough, when they arrived at the eatery they spotted Hasso Herschel with his new facial hair. With him were Gigi Spina and Wolf Schroedter, who were suspicious of the possible recruits until they learned that one had fled the East and the other had worked on the Wollank tunnel.

With that settled, Wolf picked them up in the VW van and drove them to the site. Once in the tunnel, the husky, round-faced Neumann estimated that the diggers had proceeded about forty feet. Only four hundred to go! Yet he asked, "Why the triangle shape?" In his view it was unnecessarily complicated. Just run the logs straight up and across, he advised, adding that this would enable them to put in a ceiling. Oskar, in an equally important contribution, said he could bring along five diggers from the Wollank project to help. Finally, excavating could proceed in three eight-hour shifts—around the clock.

TUNNELERS liked to boast that they were the true enforcers of the four-power division of the city—the only Berliners freely entering any and all sectors, even if underground and decidedly unwelcome in one of them. Since the escape led by the Becker brothers back in January nearly all of the attempts had gone in the West-to-East direction. There were compelling reasons for this. It was far less dangerous to start digging in the relative safety of the West, where you could also dispose of the earthen evidence fairly easily. And any tunnel organized in the East faced a far more likely chance of Stasi infiltration.

None of that, however, could stop Max Thomas. He liked to tell people he didn't even want to be buried in the East when he died—a rather pressing assertion given that he was eighty-one. Thomas had tried to join the Beckers' January expedition, but they warned that their tunnel would be too tight for Max and his elderly friends and family, who might also be prone to panic. Three months later, Thomas resolved to direct his own escape path, recruiting a fifty-seven-year-old truck driver and two others, both seventy, to do the digging. Starting from a chicken coop in the Thomas backyard, they removed four thousand buckets of dirt over sixteen days, storing it in a former horse stable.

The passage, just over one hundred feet long, was cavernous by the usual standards—about five and a half feet high. This would allow escapees to walk (a bit stooped) and not crawl, wear their best clothes, and carry suitcases—to flee with dignity. At the breakthrough site in the West the diggers constructed a ramp so that the refugees would not have to climb a steep ladder. It worked. Five women and twelve men—all but one over fifty-five—made it through on the night of May 5, emerging outdoors in a park. The West Berlin press dubbed it the "Pensioners Tunnel." Two weeks later the *New York Times* covered the episode, identifying Max Thomas as "Grandpa Fritz," who explained that his tunnel was wider than any previous "because some of us are so stout we needed more space." (A Stasi report on the episode criticized border guards for not "drawing the necessary conclusions" after the Becker escape, just four doors down the block, back in January.)

Elsewhere along the Wall, violence continued to rage. One shooting caused the biggest uproar yet. It began a little after five on the afternoon of May 23 when fourteen-year-old Wilfried Tews, in trouble back home for refusing to distribute leaflets for a Communist youth group, crept through the centuries-old Invaliden Cemetery. One shot whizzed by him as he neared the outer wall and another as he climbed over it. After scrambling down the embankment, he dove into the Humboldt Canal and began swimming the short distance—fifty feet—to the opposite shore, just as he had planned based on a tourist map he'd purchased. Guards fired dozens of shots; a bullet passed through one of his lungs, others struck an arm and a leg. Still, he managed to climb onto a platform in the West next to the canal even as he was hit again. Dozens of other shots chipped the stones nearby.

(For Tews, the "western" that came to mind at this moment was not Berlin but the Hollywood variety.) West Berlin cops, trying to reach him, returned fire. One shouted across the canal, "Stop shooting! You are German, too, aren't you?"

Then Private Peter Göring, a twenty-one-year-old East German guard, went down, struck by three rounds as he left his position to take a kill shot.

Tews would survive, but Göring perished. And so began the propaganda war. Communist officials dubbed Göring a martyr, arranged a state funeral, and started searching for streets and buildings to name after him. The SED-backed *Neues Deutschland* published a gruesome photo of Göring—a corpse on the ground, eyes staring up at the sky—on its front page. Eastern media claimed he had been lured into a trap and "assassinated." A reward of 10,000 DM (about $2500) was offered for the capture of the West Berliner who killed poor Göring. In response, West Berlin officials accused the VoPos of attempted murder, pointing out that even the guards' own rules forbade firing on women or children, such as Tews, not to mention spewing bullets into the West. The U.S. Mission observed in a cable to Washington that the Communist version of the incident omitted the fact that GDR guards fired first.

Germans trying to gun down Germans—it had been happening for months, but this high-profile episode freshly rankled citizens on both sides of the barrier. After reading about Göring's death, a woman reportedly told her husband, a sergeant in one of the GDR border brigades, "Under no circumstances should you sign up for another year's service." She was then called to the company commander's office for a "clarifying conversation." Yet the shootings continued. On May 27, another young man trying to escape was shot in the skull by a guard in a watchtower. Again West Berlin police fired back. Doctors and nurses at an adjacent hospital watched as the young victim was left in the bushes for forty minutes before he was taken away.

. . .

A second, if far less intimidating, wall was now drawing attention from West Berlin locals and tourists. The Associated Press
reported, with an accompanying photo, "Hollywood movie
makers are building a plaster wall in Berlin for a film featuring the escape of 28 refugees from Communist East Germany.
The dummy wall, about 300 yards long, is in a section of Berlin
well away from the 25-mile-long wall erected by the Communists. But to prevent any mistakes, movie officials put up notices
in four languages explaining that it is just an imitation." The
site was near the Tiergarten, Berlin's most famous park, and
the sign read, *This Is Not the Real Berlin Wall, This Wall Is for
the Movie "Tunnel 28." W. Wood Prods*. It was marketing genius.
So many tourists were stopping to see it—even with the real
Wall nearby—that they sometimes disrupted filming.

MGM hoped the movie would turn out better than the last
Hollywood feature shot in the city, Billy Wilder's *One, Two,
Three*. Wilder had the misfortune to start filming just before
the Wall went up, and had not quite finished when it emerged.
Locations in East Berlin were scrapped and the movie team
had to build a nearly life-size version of the Brandenburg Gate
out of papier-mâché near Munich. Then there was the tone of
the film—it was a comedy, following the daughter of a Coca-
Cola executive based in Berlin (James Cagney) who falls in love
with an East German Communist (Horst Buchholz). After August 1961 few were laughing about Berlin, and the film had performed dismally at the box office in both the United States and
West Germany. "What tears our hearts apart, Billy Wilder finds
funny," a Berlin newspaper lamented.

The director of *Tunnel 28*, Robert Siodmak, seemed well
chosen. He knew something about escaping oppression. Born in
1900 and raised in Dresden, and a Jew, he had fled to Paris and
then to Hollywood in the 1930s, like his Austrian friend, Billy
Wilder. After directing a few B features such as *Son of Dracula*,
he graduated to film noir, including *The Spiral Staircase*, and
earned an Oscar nomination for Best Director for *The Killers* in
1946. In the early 1950s he collaborated with Budd Schulberg on

a screenplay titled *A Stone in the River Hudson*. Deeply revised, it became the revered script for *On the Waterfront*. (Siodmak never got even a partial credit, sued the producer, and won a $100,000 judgment.) After a series of films that performed poorly, he returned to West Germany to make well-regarded dramas.

The scriptwriters for *Tunnel 28* were Peter Berneis, another German native who wrote the screenplay for *Portrait of Jennie*, and Gabrielle Upton, who had penned (improbably) the teen film *Gidget*. Producer Walter Wood hailed Siodmak's "realism," comparing it favorably to the French New Wave's overreliance on "spontaneity." The director, however, had trouble finding a single West Berliner willing to play a VoPo—some said they were afraid East German police might recognize them later and detain them at checkpoints on the autobahn to West Germany. Siodmak had to go as far as Munich to cast those roles.

According to a *Los Angeles Times* columnist, Siodmak could "scarcely contain the excitement" in a promotional call from Berlin. He had been filming a scene along a canal at the border with a crowd of West German extras, which eventually drew the attention of the VoPos, who drove three of their vehicles close to the border and shone spotlights directly into the cameras. Siodmak, who had anticipated harassment, had a second crew filming a similar scene farther down the canal. The director crowed that he not only got the two shots he wanted but "you'll actually see their searchlights playing back and forth. Talk about realism!" On another occasion, Siodmak and producer Wood stood next to the Wall on a platform and peered across it through binoculars. An American soldier rushed forward and instructed them to put down the binoculars. East German guards considered anyone using them to be spies and were known to spray rifle fire, either as warning shots or in earnest.

Erwin Becker advised the filmmakers on numerous details, from the type of soil he and his brothers had excavated to the electric cords used to connect the lighting. He visited the set most days, critiquing the actors and the atmospherics. Becker gave his blessing to the aura of grit and gloom inside the fake basement at the UFA soundstage. The plaster tunnel was quite different from his, of course: the walls could be pulled back and the ceiling raised on pulleys to accommodate cameras.

WHEN Franz Baake responded to the knock on his door at the UFA studio in Tempelhof to find three very intense young men, it came as quite a surprise. No one had made an appointment and he was busy. Baake was moonlighting as publicist for MGM's *Tunnel 28* while it was shooting in Berlin. At the same time he was putting the finishing touches on his own short film about the early months of the Wall, *Test for the West*, set to be screened at the Berlin Film Festival in June. Given the subject of both of these films, he quickly took an interest in what these visitors had to say. They were two Italians and a German, each a bit younger than the thirty-year-old Baake. One of the Italians opened with "How would you like to see a *real* tunnel?"

Tunnel 28 was starting to draw wide coverage in Germany, partly thanks to Baake's work. He had written catchphrases such as "*Tunnel 28* is not a film—it is dynamite!" When a story appeared critical of this framing, Baake asked producer Walter Wood to assure the press that MGM was just trying to help Americans and Germans understand the true evil of the Wall. Some publicity had gone international, such as the story about the replica of the Wall on the street. Among those who learned about the MGM movie through press coverage were the three real-life tunnelers.

Refusing to give Baake their names—and warning him not to talk to anyone about their project—they asked if he would take photos of their alleged tunnel for unspecified use. Intrigued, he agreed. Returning a few days later, they blindfolded him and loaded him in the VW van. Baake respected the security concerns. He knew a German journalist who had been treated like a traitor by his peers for revealing a little too much about an earlier tunnel. Baake brought along his Rolleiflex camera, which would produce images in a larger format than his usual Leica.

Emerging from the van, he was led down a few steps to the

basement of a building, and then into the tunnel. Baake, with blindfold off, marveled at the professionalism, the lights along the ceiling, and the sturdy 4-×-4 wooden supports. On the other hand, it was half dark, with puddles of water in the dirt—in his estimation, "scary." He felt his heart thumping as he crawled a few feet past the opening to snap the last of several dozen photos. Then he revealed that a friend had a darkroom he could use to produce the photos. The tunnelers insisted on accompanying him for security reasons. Baake developed the film and made a few prints. Since he found the students idealistic and supported their mission, he gave them the photos and negatives free of charge.

A few days later, the trio returned to Baake's office. Now they were setting their sights beyond simply documenting the dig with still images. They wanted to know whether Baake had any sway with American television networks who might offer the widest audience—and largest payment—for their project. Baake had no contacts, but he knew someone who did: Fritjof Meyer, who was working for a federal agency but aspired to a journalism career. This crazy tunnel project might help in that pursuit.

Meyer wanted to see the tunnel first, and so he, too, made the trip, riding blindfolded in the back of the van after meeting the two Italians and the German at a playground. Even more jittery than Baake as he crawled into the dark hole, Meyer nearly turned around when a streetcar passed overhead on Bernauer Strasse, dislodging a shower of dirt. Having survived the experience, he contacted a friend at NBC. This was Abe Ashkenasi, the local NBC part-timer who had been seeking just such a tip for weeks. Meyer knew him as an occasional drinking partner at the Eden Saloon, a favorite watering hole for Western journalists, described in the tourist guide for the U.S. military as a "bohemian type place for young people." (Another frequent patron: Siegfried Uhse.) Ashkenasi told his bureau chief, Gary Stindt, about the tunnel—and Stindt informed correspondent Piers Anderton.

It was a timely gift for Anderton. Months had passed since

his boss, Reuven Frank, had instructed him to find a tunnel. Now one had found him. And none too soon, with his rival, Daniel Schorr, pumping the Girrmann Group for access to any tunnels they knew about.

SIEGFRIED Uhse's six-month career as an informer reached new heights in May with his first escape assignment from the Girrmann Group, organized by its top American staffer. It began one afternoon at the House of the Future when he observed the manager, Bodo Köhler, in a curious conversation with a young West Berlin policeman who wanted to leave his "stuff" at the office. Köhler assented. Later Köhler told Uhse, "It's good that you came again. What we decided the night before yesterday is ready to go. The idea is so good we can hopefully put a lot into it. I cannot tell you more now, you will understand why." Uhse offered to act as courier and meet with potential escapees, aided by his new West German passport.

A few days later, Uhse met a young American at the House of the Future. Joan Glenn had come to Stuttgart the previous June as an exchange student from Stanford University (she hailed from Salem, Oregon). Visiting West Berlin that December, she happened to stay in the youth hostel section of the House of the Future. Inspired by what she witnessed, the nineteen-year-old Glenn forgot about Palo Alto and remained in Berlin to assist with the fake passport scheme while living in the basement, then the attic, of the House of the Future.

Berlin had become a kind of mecca for idealistic students from across Western Europe and the United States who wished to aid their peers trapped behind the Wall. Some, like Glenn, were studying in Germany; others devoted holidays to the cause or skipped entire semesters. They took varying levels of risk. Most stayed in West Berlin, packing the House of the Future from top to bottom. They helped Girrmann organizers match old passport photos to images of look-alikes in the East who wished to escape. Students pored over pictures laid out on massive tables at Girrmann offices. Some volunteered to take the far riskier step of carrying the photos or passports to the East, where they might be detained at the border coming or going. Several dozen foreign students had already been arrested. Two students from California and one from the Netherlands had been sentenced to prison for human smuggling, but later released in an "act of clemency" by GDR leader Walter Ulbricht. The fate of others was unknown.

Two of Joan Glenn's fellow Stanford exchange students, also staying at the House of the Future, used their U.S. passports to visit East Berlin and pass fake IDs to potential escapees. In late January one of them, Robert A. Mann, had gone missing. Glenn told Stanford officials, then Mann's parents, what he had been doing. Soon the East Germans announced that Mann had been arrested on charges of aiding the escape of a former Free University student. Mann's parents, halfway around the world in Sepulveda, California, expressed hopes he would be freed from the Brandenburg prison without a trial, but Washington could bring little pressure to bear across the border. Months later, there was still no word.

Now Joan Glenn asked Siegfried Uhse to help her smuggle a mother and daughter to the West by unidentified means. Uhse volunteered his services. Glenn confided something else: a "violent breakthrough" was planned at the border sometime between May 23 and May 25. Uhse didn't press for details, fearing that Glenn would become suspicious. But he took his source seriously. Köhler, after all, had informed him earlier that Glenn was "involved in the whole business," was in fact his "right hand." She had taken charge of keeping the vast and secret list of potential refugees up to date and organizing the couriers for future actions. And rumor had it that Köhler and Glenn were also having an affair.

SIEGFRIED Uhse did not have to wait long to find out what the "violent breakthrough" might amount to. It arrived slightly behind schedule, in the early morning hours of May 26, but the impact was, on more than one level, tremendous. The headline on the front-page *New York Times* story declared, "Four Blasts in 15 Minutes Rip Reds' Wall in Berlin." It was the most dramatic attack on the barrier yet, scattering stone and rock for hundreds of feet along Bernauer Strasse. No one was hurt—and apparently no East Germans escaped through the gap—but the blast destroyed some GDR border facilities. A West German police official said it appeared there was now "an active movement to get down" the Wall.

The *Times* reported that officials "believed underground groups of East Berliners were responsible." It published a large UPI photo of two concerned West Berlin policemen peering through a fifteen-foot hole in the Wall. Editors could not know, or possibly even imagine, that the cop on the right side of the picture was the man who had helped organize, and set off, the main blast—lighting it with a cigar, some claimed.

He was Hans-Joachim Lazai, twenty-four, long assigned to the Bernauer area. The previous August he had watched from his patrol car as Ida Siekmann jumped from her apartment to her death, becoming the first Wall fatality. He was also at the scene when a young German leaped to his death a few weeks later, missing the fire patrol's outstretched net (Lazai was among those who had encouraged the attempt). The deaths enraged him. On other occasions he felt sick when ordered to train fire hoses on young West Germans protesting the barrier. Some of his colleagues assisted escape helpers by lending them weapons or standing guard during a tunnel breakthrough. Lazai had aided several operations himself, but each had failed, and he wanted to do something even more provocative to undermine a structure he considered profoundly inhumane. To that end he volunteered his training in explosives to the Girrmann Group. (He was probably the policeman Uhse had spotted dropping off packages at the House of the Future a few days earlier.)

The leaders of the Girrmann Group were not known for supporting violence, but they passionately hated the Communist regime. The anti-authoritarianism of Detlef Girrmann and Dieter Thieme could be traced to the closing months of World War II. Still in their teens, Thieme joined the army and Girrmann applied to Hitler's *SS*. After the war, ashamed of their past, they became ardent social democrats in the East. Thieme would serve three years in prison for distributing books and flyers critical of the government, then fled to the West. By that point, Girrmann and Bodo Köhler, also threatened with arrest, had already emigrated. After years of other forms of activism, the trio had found their true calling in organizing escapes from the East.

While they opposed violence against people, an unfeeling concrete wall was another matter. With Lazai they chose a blast site in a busy, highly visible area, but ordered that no one be harmed. From a Girrmann associate, a Swiss mining student, Lazai obtained six kilos of malleable plastic explosives in twelve rolls that felt like marzipan. Fellow cops helped him unload forty-pound sandbags to be used to direct the blast eastward, through the Wall. A refugee escape plan never came to fruition—a symbolic blast would have to be enough.

Shortly after midnight on May 26, Lazai initiated the explosion at Bernauer and Schwedt. By the time it detonated sixty seconds later, he was hustling to his patrol car seven hundred feet away. Then he phoned in the report to headquarters from inside the dust-covered vehicle. Soon French and East Berlin police arrived at the scene. As the sun rose, photographers snapped pictures of West Berlin police, including an unabashed Lazai, at the site. But Lazai was not done. The next day he flew to Frankfurt, where he had arranged to pick up more explosives secretly stored at a U.S. base. Military police had been tipped off, and Lazai was arrested. West German police interrogators told him, "We don't like what you did—but we understand." He wasn't detained for long and was never charged in connection with his sabotage, merely transferred to a post in Lower Saxony.

And no wonder: He had supporters in high places. Shortly before the attack, Bodo Köhler had met with Egon Bahr, the influential aide to Mayor Willy Brandt who was known to be sympathetic to escape helpers. Bahr, who had been a newspaper reporter in East Berlin before fleeing West in protest of government censorship, despised what he called the *Scheissmauer*— wall of shit. After they discussed taking stronger measures against the East, even a detonation or two, Bahr raised his arms dramatically and said, "Something *has* to happen at the Wall! Understand?" Köhler took this as support for a bombing.

When the explosions took place, the Social Democratic Party (SPD) to which Bahr belonged was meeting in its national convention at Cologne, where Willy Brandt hoped to be elected to its number two position. Calling from a phone booth in Berlin, Köhler informed Bahr of the successful operation. Bahr's response signified to him, *Boy, it's about time*. Brandt then told the convention that students from the West had blown up part of the Wall, and he hailed them for not passively accepting its existence. The Wall, he said, was so "unnatural and inhumane that we can never accept it." This inspired a standing ovation. Brandt later won the vote for vice president of his party, setting him on a path to the chancellor's office.

ON May 27, the day after the blast along Bernauer, Piers Anderton met with Franz Baake and Fritjof Meyer to discuss their mysterious tunnel project. They gave him the address of an apartment near the TU campus where he could meet the three organizers the next day. Anderton wrote in his appointment book for May 28: *Students, etc.* This would keep the subject secret in case anyone saw this entry.

Anderton arrived to find one of the Italians—the short one, Sesta—able to converse pretty well in English. Spina, the tall one, said little. The German, Schroedter, said nothing—as he clicked the bolt on an automatic pistol. These guys meant business, Anderton decided. Sesta told him about the plight of Peter Schmidt and showed him maps of the city's underground grid and various engineering designs for the tunnel. He coolly declared that they needed $50,000 to complete the dig. That would never fly with Reuven Frank back in New York, Anderton knew, but he asked to see the tunnel. And so, like Baake and Meyer, he made the trip via VW van, with Schroedter as his escort.

By the time of Anderton's visit, the tunnel stretched more

than seventy feet, almost to the Wall. Anderton was amazed how much dirt was already piled in the cellar. The triangular shape of the wooden supports near the opening now gave way to the square design. Anderton told Schroedter he was definitely interested in filming the project but would have to get an okay, and money, from New York. By pure chance, he was about to leave for Manhattan—to get married. He would see his boss, Reuven Frank, at the wedding party. Schroedter made him sign an agreement (likely nonenforceable) that stated, *Hereby I declare that on the 28th of May 1962 . . . that I will keep silent about the enterprise of Wolfhardt Schroedter. If I break that promise, I will pay the sum of 50,000 USD to Herr Schroedter.*

Three days later, Anderton left for New York.

5
The Correspondent

Piers Anderton had brought important business to New York, but first things first: his wedding. Divorced, and the father of six, he tied the knot again in a brief civil ceremony downtown at City Hall. His attractive blond Swedish bride, Birgitta, eighteen years his junior, had seen far more of the world than Anderton, as a former Pan Am flight attendant. Piers, with a newly trimmed beard, wore a suit; Birgitta, a silk dress. NBC correspondent John Chancellor and his wife met the beaming couple outside City Hall and together they left for the network's luncheon in the Andertons' honor at the top-tier Four Seasons.

Reuven Frank was enjoying the festivities—several colleagues were getting an early start on their alcohol intake for the day—when the groom backed him into a corner and advised, "We have to talk. *In private.* In your office." Frank found this strange given the occasion, and tried to put him off, but Anderton insisted. When the party ended, instead of celebrating further with Birgitta, Anderton walked her to the curb outside and said, "I have to go to the office."

"What, you're going to leave me on the sidewalk on my wedding day?" she protested. Compromising, he joined her in a taxi to the Summit Hotel, then left to meet Frank over at 30 Rock.

After asking Frank to shut the door of his office, Anderton said, "I have a tunnel."

"What are you talking about?"

Anderton explained. In fact, not only had he seen it, he had crawled a few dozen yards inside. And he wanted NBC to pay the three tunnel organizers in exchange for exclusive access.

Frank, who had practically demanded that Anderton find such a venture the previous August, was delighted, but insisted that they keep this hush-hush. He would tell only a couple of other people at NBC on a need-to-know basis. Because phone calls out of Berlin were presumably tapped by the Russians, the West Germans, or the Americans (or all three), Anderton must communicate with him only in code or when he was traveling outside Germany. Anderton would have to act "James Bondish," as Frank put it, knowing that Piers could keep a secret: he had spent two years in the Pacific during World War II as an intelligence officer.

At this point Anderton revealed that the three tunnel organizers had demanded $50,000.

"That's crazy!" Frank said. "We can't do that."

"What about just pay for supplies?" Anderton countered.

"Well, I guess we can do that," Frank replied. But he set the limit at $7500 (still a considerable sum, as a new car in the United States was selling for about $2000) for the rights to film the rest of the digging and eventual escape, take it or leave it. Frank quickly found his boss, William McAndrew, NBC's vice president in charge of the news division since 1951. He signed off on it—though not literally, as no one was signing anything. McAndrew agreed that they should keep it from lawyers and even from *his* boss, NBC president Robert Kintner. He would find a back channel to secure the cash.

Anderton hastily arranged a pair of airline tickets and rushed to his hotel to tell Birgitta that they had to return to Europe that very night. So much for a honeymoon.

Frank knew Anderton was already deeply invested in this story, but he did not worry about his objectivity. Unlike many of his colleagues, Frank didn't consider that something documentaries should strive for. Films are made by people, not machines; the only writers and correspondents worth their salt were those deeply interested in an issue or event who then *react*. From those journalists he didn't even demand "balance," just what he called "responsibility." Frank liked to say, "You can't judge fairness with a stopwatch."

That night, Piers and Birgitta (still in the dark about all this intrigue) boarded a flight to Paris. Anderton carried the cache of NBC cash in his trousers. The couple rented a car in Paris, then drove to Bonn. Anderton left his new bride in their apartment and drove off, speeding along the autobahn into East Germany and on to Berlin.

ANDERTON knew he was hardly home free. If the three students accepted the NBC deal, and he did manage to cover their tunnel escape, there was still no telling how U.S. officials in Berlin, or those back in Washington, would respond if they found out about it. Anderton had already earned criticism and notoriety and blamed the State Department for it.

His troubles arrived earlier that spring after he reported that GDR troops were known to fire at U.S. transports on the autobahn as they brought soldiers or supplies from West Germany to Berlin along this corridor. The administration wanted to keep it quiet because it exposed the vulnerability of the American position there. A State Department official berated Anderton for revealing this. The reporter asked if his account was accurate. "Yes," the official replied, but added, "It is contrary to U.S. policy to report it." Anderton considered this an attempt to suppress the news.

An official at the Berlin Mission then (falsely) informed Secretary of State Rusk that Anderton seemed to hold a "very negative 'US should get out of Berlin' line," which made them reluctant to cooperate with him. Rusk replied that the "situation in the corridor should be played quietly." While newsmen should be allowed some access, "we do not feel it wise to go out of our way to provide extra facilities or in other ways build up publicity or public treatment of situation or otherwise greatly cooperate." A few days later, Under Secretary of State George

Ball warned the Bonn embassy that Anderton's reporting tended toward "Voice of Doom" and "has been highly critical" of government policies.

Anderton drew more criticism, this time very much in public, following a twenty-minute speech in April to a women's group in Germany made up largely of wives of American diplomats and other officials. Alleged extracts from the talk were leaked to *Variety*, which detailed them in a front-page story titled "NBC's Anderton's Incendiary Berlin Talk Shocks Wives of U.S. VIPs." He had apparently attacked Kennedy and Rusk for being (in *Variety*'s words) "wishy-washy" on Berlin. Berliners were "two-faced" for privately complaining about dangers but urging the press to downplay them so they would not lose tourist dollars. As for the American public—they didn't know or care much about Berlin. All of this from Anderton had caused the wife of U.S. Ambassador Walter Dowling to leave the room; other club women were "shocked." Anderton had even ripped his own network, accusing NBC of attempting to "muzzle" (again, *Variety*'s term) its correspondents. This, most of all, was sure to rankle his superiors. As if to ensure this, the article's opening sentence asserted that, because of his remarks, he was now "in hot water" with his bosses back in New York and with "the top U.S. government representatives in Germany."

Anderton was livid. Quotes had been yanked out of context or maliciously mischaracterized. In some cases his opinion was the opposite of *Variety*'s summation. He had not claimed that NBC was muzzling anyone. The whole episode was, it seemed to him, clearly a State Department hit job to force the network to reassign or fire him. So far his bosses had backed him, but now in June he consulted a lawyer about suing *Variety* for libel. Meanwhile, two reporters told him they had seen a State Department cable charging that he was "pro-Communist."

AS the Bernauer tunnel neared the border and death strip, Harry Seidel was hardly idle. He and Fritz Wagner had cooked up a promising project to defeat the Wall back at their old digging grounds. *Dicke* had paid 4000 DM (about $1000) to the owner of the Krug pub at the corner of Heidelberger Strasse and Elsen Strasse for the temporary use of his cellar. Their target was only eighty feet away on the other side of the border: the basement of a photo shop, which would be closed for the Pentecost weekend.

Heidelberger remained perhaps the most bizarre residential address in the world. The GDR border actually extended from one side of the street to the other, to the façades of buildings in the West, but unlike at Bernauer Strasse the East Germans could not brick up all the doors and windows on that side. Instead they built the Wall down the center of the block, even though half the street and the sidewalk across the way were still on East German territory. West Berliners could enter and exit their tenements, and children still played on the sidewalk, but it was GDR territory and closely scrutinized by VoPos. If a West Berliner parked his car there for too long a spell, it was liable to be claimed by the Communists and hoisted right over the Wall.

Led by Seidel, two teams of three men started shoveling on June 6, taking twelve-hour shifts, then resting on mattresses under the pub. Food and drink had been stored there for a weeklong immersion. A West Berlin police detective who moonlighted as a Stasi informant nosed around upstairs, but failed to notice anything amiss. A Stasi report called the pub a "conspirative nest" frequented mainly by West Berlin cops looking to get drunk and young people plotting "provocations." One such troublemaker who lived above the pub liked to haul his stereo speakers to the balcony and blast speeches by Western leaders over the border to the East.

Guard patrols and inspections by Stasi operatives had been stepped up due to past tunneling in the area, but the diggers were undeterred. One of them was young Peter Scholz, a butcher who was aiming to evacuate his fiancée and her four-month-old baby. Like many of the others, he was awestruck to be working with "*the* Harry Seidel," as he was often referred to, due to his fame as a cyclist. Harry barely seemed to rest; he was known to grab a shovel from a man who was slowing down and take the rest of his shift. To find out if the tunnel, which was just a hollow shell with no supports, would likely hold, Harry ordered someone to drive over a section of Heidelberger in a truck loaded with coal. When only a little dirt fell from the ceiling he figured it was safe—or safe enough. Even though Seidel seemed like the strong, silent type, some diggers had seen him lose his temper. One time Wagner's constant chatter and boasting so irked him that he threw a shovel at the fat man, shouting, "For goodness sake, shut up, *Dicke*!"

Above ground on Heidelberger, on the evening of June 8, several West Germans tossed bottles and stones over the Wall at border guards, who were forced to take cover. Then a ladder was pushed against the Wall in the East and a young woman started climbing over. Three shots rang out but she made it to the other side. A Stasi informer reported that the legendary Harry Seidel was spotted near the scene twenty minutes later.

A few days later, Harry broke through the photo shop's cellar wall in the East. The following afternoon, a Sunday, waiting in the photo shop with curtains drawn, the diggers tried to relax, though they knew East German police were patrolling just yards away. Several were armed with pistols. Earlier that day Wagner's courier, Dieter Gengelbach, had told the refugees to come to the building in staggered intervals. Two women soon arrived at the door, partly hidden under a porch, burdened by large packages, contrary to instructions. One of the women, despite knowing she'd be crawling through dirt and mud, wore a fur coat. Harry quietly expressed his disgust as he escorted her into the tunnel.

Two more women arrived with children. One of the women began to panic and fell to her knees in prayer. Peter Scholz lifted her off the floor and guided her down the stairs. When she began to shriek, he covered her mouth with his hand. But he was near panic himself. His fiancée, Erika, and her baby daughter had not arrived and time was growing short. After asking Harry for permission, he phoned her at the hospital where she worked. Erika hadn't heard from a courier. As instructed, she gave the baby part of a sleeping pill and hailed a taxi to the photo shop. Peter took a chance and went outside to meet them. Seeing this, Wagner muttered, "Bloody fool!" When they entered the shop, Erika handed her sleeping daughter to Harry Seidel, who placed her in one of the large tin pots they used to transport dirt—and dragged the bowl through the tunnel. Soon they were in the pub's cellar, where Seidel handed Erika the baby and said, "Welcome to West Berlin."

The following day, the *New York Times* reported that twelve in all had come through the tunnel (the actual total may have been more than twenty). "East German officials," it related, "are said to be alarmed at the growing rate of escapes and the continuing unrest along the border." When Harry Seidel returned home after this success, his wife begged him to make this his last tunnel. He could be proud about what he had accomplished already and retire on a winning note. Harry refused yet again. He still had to free his mother. And he had already settled on the location for his next tunnel, returning to the area he knew best, where he had helped his wife and son and so many others slip through the barbed wire: Kiefholz Strasse.

IN the early days of June, sad reality confronted anyone who hoped that violence along the Wall might decline after the shooting of teenager Wilfried Tews and the death of border guard Peter Göring. If anything, desperation in the East was growing, leading to an upsurge in escape attempts. "The wall has not fulfilled the Communists' hope that it would stabilize the situation in East Germany," the *New York Times* opined. "Instead, discontent and restiveness are said to be mounting among East Berliners. Dramatic flights . . . have occurred almost daily the last few weeks."

Apparently in response, East German and Soviet work crews raised the height of the concrete wall in some places and constructed a second, inner wall on the far side of the death strip in others. They installed tall, wooden screens so that Berliners could not even wave to friends or relatives across the Wall. Police patrols were increased, and more guard towers, with better sight lines, were erected every week. Land mines were planted strategically in spots. To prevent escapes by water, barbed wire was installed in canal beds and along gates strung at midchannel. Still, according to West Berlin police, eighty-six refugees conquered the Wall in June—mainly slipping or snipping through wire and fencing in outlying districts—including six members of the Soviet military.

Police also reported no less than nineteen new cases of East German guards firing on refugees. It began when a sixteen-year-old West Berliner caught a bullet in the pelvis trying to help a friend escape. Then Axel Hannemann, seventeen, was shot dead in the Spree after leaving this note for his family: "I have no other choice. I'll explain my reasons to you when I have made it. But for now I can say that I have done nothing wrong." Two days later, a pair of East Berlin teens tried to scale the cemetery wall that joined the barrier at Bernauer Strasse. One was shot in the leg, but both made it to the West.

A particularly cinematic escape was engineered by a group of fourteen East Berliners who chartered an excursion boat for a cruise on the Landwehr canal. A party broke out and the revelers, as planned, got the captain and engineer drunk. Then one of the passengers took the wheel and guided the boat West, as machine-gun bullets fired from the East clanged off the heavy metal wheelhouse where all (including one infant) had gathered for safety. The barrage continued even after they docked in West Berlin. More than two hundred shots were fired, with some striking houses and buildings in the American sector. One bullet went through the window of a cafeteria; no one was hurt. West Berlin police returned fire. The refugees disembarked, and the ship's captain and engineer were allowed to pilot the vessel back to the East. Escapees told reporters they were so desperate they would have jumped off the deck and tried to swim to the West, even through machine-gun fire, if their seizure of the ship had failed.

Elsewhere a small tunnel collapsed and buried an escapee. Fortunately this occurred after he had reached West Berlin. Three other refugees dug him out using the shovels and soup-spoons they had used to dig the tunnel.

Then another incident ended with gunfire—this time from a tunneler's weapon. The clash was triggered one evening when a VoPo patrol noticed a man, a woman, and two children head-ing for a four-story building in central Berlin. Police had grown suspicious earlier that day after spotting cameras on the roof of the Axel Springer building, which bordered the Wall in the West. When they asked the suspects for their IDs, the woman and children started to run as the man, Rudolf Müller, pulled out a pistol, shot one of the guards, and then joined the others in scrambling to the West. Müller, the woman's husband and father of the two children, had dug the tunnel with his brothers and friends, starting from the Axel Springer complex.

A reporter captured the chatter of the large crowd gath-ered on the Western side. One student who boasted that he had worked on the tunnel said of the American army, "They come here, chewing gum, and do nothing."

"Nobody does anything to help the poor devils over there."

"What can the Allies do?"

"Nothing, nothing. Germans are shooting Germans, isn't that enough?"

The guard, twenty-year-old Reinhold Huhn, died and, like Peter Göring, was immediately transformed into a Communist martyr. East German officials called for the arrest of the shooter. Müller first denied, then admitted he had a gun, but told West Berlin authorities that Huhn had been shot by accident by one of his fellow guards. Local authorities and media in the West (including CBS's Daniel Schorr) repeated the lie. And so another propaganda battle raged for days.

It was clear that, as the CIA's "Current Intelligence Weekly Summary" had recently predicted, "[w]armer weather and the summer vacation season will probably bring an increase in in-cidents along West Berlin's sector." Six refugees had died near the border in the past month alone, bringing the post-Wall total to at least thirty. "West Berlin leaders are already alarmed by the number of escape incidents, the frequency and seriousness of shootings, and the efforts to destroy the wall with explosive charges," the CIA report revealed. By whatever means, the spy agency seemed to possess accurate information on escape operations:

> Many East Germans and East Berliners flee with the direct assistance of West Berliners, chiefly university students. When alerted in advance, West Berlin police hide near the border to help if needed. . . . Tunnels have become a common means of escape. West Berlin students . . . apparently have made use of city planning maps and first-hand knowledge of the city's streets, elevated train lines, and sewer system to plot excavations from buildings immediately adjacent to the border into nearby East Berlin buildings. They succeed in contacting prospective refugees—many of whom are former students or relatives—and sometimes send one of their number into East Berlin to guide the escapees.

Mayor Brandt, meanwhile, had pledged that his police would use their weapons to assist those who fled. West Berlin had countered the East's increasing militarization at the Wall by building its own watchtowers and equipping police cars with new weapons, such as M-2 carbines. Berlin was coming to a boil as summer neared.

PIERS Anderton, it turned out, didn't have to play hardball to get the Bernauer tunnel organizers to accept NBC's offer of $7500 for the rights to film their adventure. Fritjof Meyer negotiated for Spina, Sesta, and Schroedter, with hopes that he and his friend Franz Baake would be involved in the NBC production. In the end, however, the tunnelers opted to cut them out of the contract (and any chance to help with the filming). Working with just NBC would simplify not only the project but also the status of their rights and payments down the line. This meant, however, that at least two outsiders in West Berlin knew about their tunnel, even if Baake and Meyer could not identify precisely where it was located. Mimmo Sesta warned them that if they ever spoke a word about it they would "have to deal with our whole group."

The NBC contract, signed by the three organizers on June 17, promised a bonus of another $5000, to share, if and when the film was completed. That would bring the total payments to $12,500. Spina, Sesta, and Schroedter would each receive a copy of the footage and retain the right to sell any photos they took. If NBC blew the secret of the tunnel before escape day, the network would have to pay the trio another $50,000. Reuven Frank never did show the contract to NBC's legal department, so it was unclear whether it would stand up in a court of law. For the organizers, any qualms about accepting payment were rationalized away: the upfront money might be spent entirely on equipment, food, and other supplies, while the cash at the end would be compensation for the dangers they'd faced and the college classes they'd missed. The bonus money also provided incentive for completing the operation as quickly and safely as possible.

One reason the tunnel organizers accepted much less from NBC than they had requested was that they had just found a way to replenish their supply of wood at virtually no cost. One of the diggers had a wealthy uncle, Dietrich Bahner, a fierce anti-Communist who had fled the East and was now active in Free Democratic Party politics in Bavaria. He often told his children, "We ran away once, we can't run a second time!" Among Bahner's business properties was a lumberyard in Wellenburg, and his son, Christian, said he would donate as much wood as the tunnelers needed. There was just one catch: Christian, an economics student at a Berlin university, desperately wanted to accompany the diggers when they broke through into the East. This seemed a bit risky—as much for the tunnelers as for young Bahner—but organizers accepted it. They would worry about that later.

Another sensitive issue loomed larger: What should they tell the other diggers about the NBC filming—and the NBC money? The three felt they deserved compensation, but they also knew that this was supposed to be a purely idealistic venture. An organizer pocketing any payment would surely rub some the wrong way. But if they didn't tell anyone, how long could they keep this a secret? NBC's camera crew was due to start shooting soon. The only way to keep the deal hidden was to make sure that the three of them were the only ones on the shift whenever NBC showed up.

The digging, in any event, was now going smoothly, although they had to hope their path to the basement of the Bulgarian's tenement on Rheinsberger in the East was accurate. (They were still working on obtaining surveying tools.) When they reached what they believed was the borderline, someone posted a sign overhead mimicking the much photographed one at Checkpoint Charlie: *Achtung! You are now leaving the American sector!* It didn't matter that, until then, they had actually been digging under the French sector. Soon the tunnelers were directly below the barren death strip, with its ceaseless patrols, sniper nests, and guard dogs. Any loud noise would carry readily above. Guards who might hear them, whether by chance or through listening devices pressed to the ground, could drill a hole and then drop a dynamite charge (as they were known to do).

By now the young men had reached a point they knew was coming: air was running thin at the digging end of the tunnel. They had used bulky industrial fans to direct air inward but that was now useless. Undaunted, two of the engineers concocted an elaborate system involving dozens of three-foot stovepipes taped together, extending the length of the shaft attached to the sides or ceiling, with motors from vacuum cleaners or a ventilator blowing oxygen through them. It worked. For fun one day a tunneler poured a few drops of perfume into the piping; on another occasion, cognac. The scents wafted to the Eastern end of the tunnel, reminding the diggers of what they were missing back in the real world (or was *this* now the real world?).

Joachim Rudolph continued to showcase his expertise with

electricity, adding to the long row of lights and hooking up a motor and a new winch to their primitive rail system to speed the dirt cart along its way. Phone cables were lengthened. The diggers joked, "This is now the only telephone line between West and East Berlin!" Even if they had to crank the phone to use it.

Uli Pfeifer, one of the few tunnelers who held down a steady job above ground, as a construction engineer, took shifts only on Friday nights and weekends. He struggled with the lack of free time and the severe contrast between his above- and below-ground lives. But he volunteered his mother's house in Charlottenburg for some of the Saturday meetings, when a dozen or more diggers met. His mother did not remain clueless about these confabs for long. Her son's unusually dirty laundry had already tipped her off.

There was a changing room on the first floor of the swizzle stick factory, its windows covered. Mud-caked shoes rested on the floor next to piles of dirty clothes to be washed (on- and off-site), under ropes holding dozens of shirts and pants, clean and soiled. The daily routine for most of the diggers was: Catch a ride from Wolf Schroedter to the site—or walk or take a bus. Arrive at the factory carrying a lunch pail or bag of food for a meal break. (Just another blue-collar worker in the city.) Swap fresh clothes for grungy ones, and work for eight hours with two or three others on the shift. You might dig for two hours, then switch jobs and yank the cart to the rear and discard the dirt. If another tunneler shared dumping duties with you it wasn't so bad—you had someone to talk to, at least. There was running water in the cellar and a large sink to wash hands and clothes. Also an open sewer pipe, about eleven inches across, where one could get rid of dirty water, urinate (when not using the piles of dirt for that purpose), or defecate. Another room held a mattress or two where diggers could nap.

Anderton on his visits sometimes crawled deep into the tunnel himself. Once, while staying with his wife at the historic Kempinski Hotel in West Berlin, he slipped out in the middle of the night to visit the site. Still in the dark about the secret operation, Birgitta awoke to find Piers returning to their room wearing mud-caked boots.

"Where have you been?" she asked, switching on the light.

"Oh, just out on another story," he said, ending the interrogation right there.

After more days passed, the tunnel engineers knew they had to make sure they were taking the right route. It was obvious that the tunnel snaked a bit to the left here, to the right there; it was tough to eyeball a straight passage while lying on one's back while shoveling. And sometimes they had to curve around a boulder. Finally they got permission to borrow surveying instruments from a lab at TU. No one had ever used these underground before, and there were no fixed points or landmarks to serve as a guide. When they came up with the final reckoning they were shocked, and delighted, to find that somehow their "snake" was still pointing nearly perfectly at their target in the East.

One more task needed to be completed: acquiring additional firearms for protection in the months ahead. Mimmo Sesta got a tip on someone in Hamburg who might provide the illegal weapons. Gun control there was far less stringent than in Berlin, and it was a mecca for youth. (An up-and-coming group from England called the Beatles was drawing large crowds at the Star-Club.) Once in Hamburg, however, Mimmo found he didn't trust the arms dealer, and ended up buying only a cheap hunting rifle.

WITH the contract signed and the cash delivered, NBC was finally ready to start filming. Back in New York, Reuven Frank had approved the final deal but still felt worried, even conflicted. Yes, the diggers were volunteers, but the risks were enormous and perhaps as a seasoned adult he was wrong to encourage (subsidize, in fact) the young and the foolish. He wondered whether he should have acted more like the British captain played by Errol Flynn in the movie *The Dawn Patrol*, who protested letting his airmen fly rickety old planes in World War I—telling the major (Basil Rathbone), "You can't send those kids up in those crates!"

Frank had mandated a few ground rules. Piers Anderton could not offer any advice or physical support for the diggers; Frank didn't want NBC accused of aiding and abetting, even if its hard cash already was doing that. The correspondent could use only two technical people in the tunnel, a pair of Bavarian brothers who did a lot of work for NBC: Peter Dehmel, twenty-eight, would shoot the footage and Klaus Dehmel, three years younger, would handle the lighting. Only black-and-white film would be used. No one else in the West could know about the filming, except for Gary Stindt, NBC's Berlin bureau chief, and maybe one camera operator above ground. To further bolster security, Frank would never visit the tunnel and not even be told where it was located until the final days. Anderton should never call or write him with progress reports. When necessary they would meet in Paris or London.

June 20 would be the momentous day. Since they had never met and did not necessarily trust the Dehmel brothers, the tunnelers blindfolded them en route to the swizzle stick factory. Once inside it was easy enough for the NBC pair to capture diggers in the cellar dumping dirt, sawing wood, or just relaxing, despite the dark and primitive conditions. But how to film inside the tunnel itself, where they could not stand to shoot film or aim the lights properly? The Dehmels improvised. Peter would lie on his back in front, feet facing forward, and point the camera (wrapped in plastic to deter moisture) into the shaft while Klaus lay on his stomach and directed the battery-powered lights from just behind him. This was extremely messy and uncomfortable, and they could shoot for no more than two and a half minutes at a time. In this confined space they had to use the smallest professional camera, and that was all the raw footage its film cartridge could hold.

UNLIKE his chief competitor, Daniel Schorr of CBS had not yet found a tunnel. Desperately seeking a major scoop to mark the first anniversary of the Wall on August 13, he put out word again to his contacts with the Girrmann Group. Tunnels were now making news regularly. Why couldn't he find one? Maybe it was because he was based in Bonn; Piers Anderton and most print reporters worked out of Berlin. All Dan could do was keep trying. Something was bound to break soon.

In the meantime, he had plenty to offer CBS News. His June 15 report opened, "The barricade around the ninety-five-mile perimeter of West Berlin, now ten months old, is being feverishly reinforced." It was becoming what Allied officials called an armed, fortified national frontier, with new barbed wire, firing slits in the Wall, and concrete shelters. "The reason, obviously," Schorr explained, "is not fear of any Western attack, but to facilitate shooting down fleeing refugees with relative security from West German police counter fire." Mayor Willy Brandt had met with his administrators that day to consider what, if anything, to do about it.

Three days later, in reporting on the shooting of Reinhold Huhn, the East German guard, Schorr quoted Brandt declaring at a rally before 160,000 people: "Every one of our policemen, and every Berliner should know that the mayor stands behind him when he does his duty, when he defends himself and when he gives all possible protection to his persecuted compatriots." Schorr added his own commentary: "West Berlin police are under Allied orders to fire only in self-defense. Other Berliners make their own orders. And, for many, their aim is to rescue as many as possible from behind the Wall."

On June 20, Schorr relayed the ominous news that on the eve of Secretary of State Rusk's first visit to Berlin, police were stacking sandbags and adding earthen screens on the Western side of the Wall for protection in case of gun clashes. "It's understood that Mayor Brandt will tell Secretary Rusk tomorrow," Schorr intoned, "that tensions will continue as long as the Wall stays up, and that West Berlin could not go along with any policy based on recognition of this barrier." While Rusk feared provoking the Russians into an explosive response at the Wall, Brandt would emphasize that as mayor "he could not order his police to stand by while refugees are shot down without a catastrophic effect on West Berlin morale."

AS it happened, Dean Rusk arrived the day after NBC started its subterranean filming. Rusk, fifty-three, and a native of Georgia, had not been JFK's first choice for this position. Kennedy had wanted the tough-minded Arkansas senator J. William Fulbright but was finally convinced that pick would prove controversial (for one thing, Fulbright was a segregationist). Rusk did not have many strong backers, but few opposed him—he had made few waves as director of the Rockefeller Foundation.

Since taking control of State and its six thousand foreign service officers, Rusk had lived up to his reputation as intelligent, cautious, diplomatic. He had taken a neutral stance on Kennedy's disastrous Bay of Pigs decision and was not yet sold on an American buildup in Vietnam. He was a strong anti-Communist, to be sure, but unlike some of his colleagues he hated "brinksmanship" and favored negotiations with Khrushchev. Some JFK aides and Allied officials felt that Rusk was far too soft in his talks with the Soviets, too eager to yield. Rusk's reluctance to take strong or original positions annoyed the President, and the two did not have a warm relationship. Some in the White House called him "the Buddha" for his inscrutable manner and bald head. Rusk was annoyingly buttoned-down in the often casual "New Frontier," sometimes wearing a suit on the presidential yacht. He even referred to himself as a "square." Rusk talked directly with the President several times a week but grew unhappy that many "eyes only" cables sent to him, including those from Germany, were often forwarded to Kennedy.

JFK's chief foreign policy advisers remained his brother Robert Kennedy, national security strategist McGeorge Bundy, and Secretary of Defense Robert McNamara. No one in this inner circle was any kind of expert on Berlin, though, so on that subject Kennedy had essentially been acting as his own secretary of state. Rusk, as was his wont, had proposed no radical moves on Berlin. In fact, he once confessed, he tried not to think about Berlin when he went to bed. He considered the division of the city fundamentally irrational, impossible to fix—he'd be happy enough if tensions did not get worse. Kennedy, by contrast, was increasingly obsessed with Berlin. At least one aide felt he was "imprisoned" by it.

Still, in June 1962 it was Rusk, not Kennedy, who visited the city. Rusk had harbored mixed feelings about the German populace since visiting the country as a student during the Nazi era (which could do that to you). Later he served as a Pentagon staff officer involved in the occupation of Germany. He wasn't sure the West Berliners would survive another Soviet blockade—he felt they were "nervous as cats" and "biting their fingernails" after the Wall went up, as if this was an overreaction. Now, riding in the motorcade during his June visit, with crowds up to fifteen thousand greeting him, Rusk wondered aloud how many of them had been "cheering Hitler" back in the day.

Mounting a viewing platform at Potsdamer Platz with Mayor Brandt, Rusk declared that the Wall "has to be seen to be believed. It is an affront to human dignity. The Wall will be broken eventually. That is the story of human freedom." He took pains at other stops in Berlin, including at Bernauer Strasse—as tunnelers shoveled dirt into a cart almost directly below—to vow that American support remained stalwart. "We Americans are shoulder-to-shoulder with you," Rusk said, "for the sake of our own freedom."

ON Siegfried Uhse's latest visit to the House of the Future, Bodo Köhler informed him that the Girrmann Group was changing its policy. Couriers like Uhse would now be assigned to a single group of East Berliners plotting to escape rather than disparate ones. His new cell would include two teachers, two high school students, a young woman, and her mother. Uhse was ordered to deliver coded typewritten messages to them. The method of smuggling the notes through the checkpoints? In the tubing of

a "telescopic" umbrella. Two days later Uhse picked up his first
loaded umbrella at the House of the Future. Before delivering
the note to an East Berlin dissident he naturally passed it to the
Stasi. An MfS agent copied the note and photographed the tricky
umbrella, opened and closed.

When Uhse returned the umbrella that day, Köhler acted
aloof and declined to speak to him. Uhse wondered if Köhler
might be on to him. But the normally businesslike American
girl, Joan Glenn, acted quite friendly when they chatted in
the kitchen. Glenn was slender with straight brown hair, large
eyes, an oval face, and a light voice. She dressed well, usually in
slacks, and spoke nearly perfect German. Köhler had arranged
a stipend for her from the Technical University, where she was
studying the philosophy and history of religions.

A few days later Köhler told Uhse that he would no longer
get his assignments in person at the House of the Future but
over the phone or off-site. Joan Glenn would now be his chief
contact. When Uhse asked for a reason, he was told to "read
the newspapers." That's when he learned about the latest fatal
shooting in a tunnel.

It grew out of a tidy three-man operation, with two of the
men motivated to the highest degree—they had been separated
from their wives since the Wall went up. One of them, Siegfried Noffke, had seen nothing of his newborn son save for a
few peeks over a border fence. Noffke, a trained bricklayer who
had worked as a driver since arriving in the West a few years
back, had found the starting point for the tunnel in the basement apartment of a locksmith on Sebastian Strasse, across from
the Mitte district. Work had gone well for several weeks on the
fairly short (hundred-foot) passage.

Then, misfortune: Ernst-Jürgen Hennig, the brother-in-law
of one of Noffke's diggers, happened to be a Stasi informer in
West Berlin. "Pankow," as Hennig was code-named, apparently
had few qualms about putting his own sister in danger of arrest. Thanks to him, the Stasi launched a mission they called
Maulwürfe ("Moles"), spying on more than a dozen would-be

refugees for three weeks, with plans to arrest them and the diggers once the project came to fruition. Hennig easily infiltrated the group. He even helped select the breakthrough point. Reaching the cellar in the East on the afternoon of June 28, he joined Noffke in chipping through the wall.

When the tunnelers stepped into the cellar in the East, unarmed, Stasi agents were waiting. The plan was to arrest the three diggers, but one MfS operative got rattled and opened fire. Noffke was critically injured. While he lay dying, the Stasi attempted to wring a confession out of him. Another digger was badly wounded and arrested. The third escaped. Hennig's sister and Noffke's wife were arrested along with the other refugees. The informer would receive a cash reward from the Stasi.

Two days later West Berliners laid a wreath on a nearby street. The ribbon read, *To our dear friend Siegfried Noffke as a final farewell from his friends. He died a victim of the Wall.* East German officials expressed no remorse in press accounts. The Stasi had once again prevented the invasion of "provocateurs," "terrorists," and "armed agents" from the West. Moreover, GDR officials claimed, they knew about five other tunnels that would soon meet the same fate.

6
The Leaks

Every other week the digging schedule for the Bernauer tunnel was established at the Saturday group meeting, with Gigi Spina writing it down to post on sheets of paper in the factory basement. The schedule for one week in early July listed three eight-hour shifts each day, including the weekend, with teams of three or occasionally four. For Tuesday it read: *Mimmo, Wolf, Kleiner, Jurgen / Rainer, Gunther II, Langer / Rudolf, Achim, Hasso.* Every so often, Piers Anderton would contact the three tunnel organizers and set a time for the Dehmels to stop by. Spina, Sesta, and Schroedter made sure only they were working on these shifts.

The Dehmels tried to capture every detail: the first-aid kit, the buzzer that signaled a cartload of dirt was ready to be retrieved, the dirty clothes hanging on lines upstairs. Young men, sometimes shirtless, were shown shoveling, studying maps, talking on old Army phones, eating sandwiches, napping on an air mattress, and inserting wood along the sides and roof of the tunnel. The camera might introduce a section by focusing on the handwritten card *Stindt-Anderton Production—Refugee Special.* Both of the Dehmels were engaged to be married, and they had a lot of explaining to do to their girlfriends as to why they were tied up for hours late at night at a secret location.

For some of the tunnelers, Anderton found himself serving as a confidant, even a kind of exotic father figure, with his gray and white hair and tales of his travels as a foreign correspondent. He had to constantly remind himself of Reuven Frank's orders not to offer any advice or assistance. When possible he tried to shift discussions from the practical to the political or philosophical.

The shaft under Bernauer Strasse was always damp and humid. Anderton felt it had the musky smell of centuries past, undisturbed until now. Water dripped from condensation on the wall or from dampness in the soil above. The tunnel crew continued placing wood planks on either side of the rail line down the center. Thanks to their lumberyard benefactor, they now had an ample supply: several tons of spruce cut into four different sizes before delivery. When they received the invoice they found that, as promised, there was no charge for either lumber or delivery, just a laughably nominal sum for weighing the cargo: 6 DM. A buck fifty. They would still have to deal with their benefactor's son, who was unlikely to forget the promise that he must accompany the diggers when they entered the East Berlin basement. He was already making them nervous. On one of his visits to the swizzle stick factory he had eagerly displayed a machine gun and a .38 Smith & Wesson revolver.

Arranging the work shifts so the Dehmels could film just the three tunnel leaders soon became a chore. The organizers decided to tell two more recruits—Orlando Casola and Joachim Rudolph—about the deal. Only Rudolph would be offered any money, however. He was prized because NBC wanted to film the electrical/technical wizard whenever possible. He would get 1000 DM up front, with the promise of 1000 more after the escape. Upon learning about NBC's role, Rudolph raised security concerns, but the organizers assured him that the TV visitors had been sworn to secrecy. The network had a lot to lose (including money) if their filming put the project at risk.

Five tunnelers now knew about the NBC project. Another dozen did not, including Hasso Herschel and core engineers Uli

Pfeifer and Joachim Neumann. But as the tunnel extended over 160 feet and well into the East, personnel worries receded. Instead, the organizers faced their first true crisis.

A persistent drip from the ceiling of the tunnel close to the border under Bernauer had turned into a full-fledged leak. Water pooled on the dirt floor an inch deep, several inches in some places. Through contacts with the West Berlin fire department— always ready to help escape attempts, going back to the days of catching window jumpers in their nets—the tunnelers obtained a hand pump and a hundred yards of hose to remove the water. It flushed into a pipe, leading to a sewer line that carried it (to their amusement) deep into the East. As hard as they pumped, evacuating about 8000 gallons of water the first week, they couldn't keep up with the leak. So they resorted to an electric pump borrowed from the firemen, but still the water rose.

Some sections of the soaking-wet walls and ceiling fell. Tunnelers discovered firsthand a problem with clay: solid when dry, it turns quickly to mud with just a modest amount of moisture. In some places the mud looked and felt like butter; in others, like black soap. The diggers' clothes, shoes, and equipment were caked in more gunk than ever. The Dehmels captured all of this in their short bursts of footage. Nearly two months of exhausting and dangerous digging had come to this: aborting the project if a fix could not be found. They hoped unusually heavy showers were to blame, but as the rain stopped the leak grew worse. The culprit had to be a burst pipe.

This conclusion, at least, produced a new idea: since the leak was centered under an area still in the West, perhaps they could get the city's water department to come to the rescue. Sesta and Spina visited a director of the utility on the pretext of interviewing him for a school paper. The director saw right through this. He told them he'd be happy to help but they would first have to clear it with German and American intelligence agencies. From the Girrmann Group, Wolf Schroedter got the name of the man they needed to see.

Known to activists only by the pseudonym "Mertens," he was an operative for the Landesamt für Verfassungsschutz (State Office for the Protection of the Constitution), or LfV. This was the West Berlin intelligence agency for domestic security. Mertens was reputedly in charge of monitoring all local escape projects. The LfV had assigned to him the dual task of keeping his government, and therefore the Americans, informed about rescue work, while also serving as a contact point when *fluchthelfer* needed official assistance or a way to reach others of their kind. In return, escape organizers agreed to notify Mertens whenever refugees they aided arrived in the West, so the LfV could interview them first.

This required considerable trust. The Stasi had penetrated West German and West Berlin government agencies, and the tunnel organizers weren't in the habit of readily trusting anyone. Mertens, a hefty man around forty years old, who combed his thinning light brown hair straight back, might have loose lips; he might even be Stasi himself. Still, his heart seemed to be in the right place. Early in the year he had warned diggers that the Stasi knew about one of the Heidelberger tunnels, and it was safely abandoned. And it was Mertens who first put Bodo Köhler in contact with Girrmann and Thieme. He even played cards with Thieme on a regular basis. For months he had assured escape helpers that he was a semi-independent operative, free to keep what he knew from his superiors so long as he prevented episodes that might hike tensions along the border. (The fact that he had to file regular reports with the CIA never came up.)

Now, at an office on Ernst-Reuter Platz, the tunnelers told him about their project but not the precise location of its home base nor its path. He thanked them but said that in this case they had to talk to the Americans directly. So Spina and Sesta trotted off to what was known in the press as "P9"—a villa at 9 Podbielski Allee.

Much of the U.S. spying activity in Berlin was based here, in the suburban Dahlem section. Just down the block from the villa housing the CIA office was America's Armed Forces Network Berlin, as well as the U.S.–German radio and TV station,

RIAS. The latter boasted the biggest transmitter in Europe. Another villa served as temporary quarters for the American ambassador when he visited from Bonn. If refugees from the East were found to have especially useful information upon arriving at Marienfelde they would be ordered to report to the villa at 9 Podbielski.

U.S. intelligence officers allegedly weren't looking to cramp anyone's style, but they did want to keep up with any escape initiatives so they could (1) stay out of the way, (2) respond if something went terribly wrong, (3) maintain lists of those involved for . . . whatever reason. Most escape helpers felt the Americans were far more opposed to tunnels than their French and British counterparts. On this day in July, however, Spina and Sesta received a friendly reception from the Americans but were asked to provide the name, address, and age of everyone working on their project. Since the Americans were, as far as they knew, on their side, the Italians provided the information willingly. The Americans assured them that they would not stand in the way of their project and might even offer support, without revealing exactly how or what.

Soon enough the West Berlin water department was in fix-up mode. Repairs along Bernauer Strasse had to be done in a casual manner to avoid alerting East Berlin guards that this was anything but routine. As a handful of city workers broke through pavement to inspect the pipes, Harry Thoess, who had worked as an NBC cameraman in Germany for more than a decade, filmed the activity from upstairs in the swizzle stick factory. Uli Pfeifer went out to chat with the city workers. He saw that the lead pipe, probably dating to the previous century, was located only a yard or so under the sidewalk. Pfeifer surmised that the roof of their escape tunnel had been dug a little too high, loosening the earthen support for the pipe and causing a crack. He had to laugh when the workers said they had never seen anything like this: Why wasn't the earth directly under the pipe more muddy? Where was all the water from this leak? The workers didn't know that it had dripped into a secret chamber below.

Repairs, in any event, were soon completed. The tunnelers

held their breath. The leak slowed, then stopped. They were good to go.

But not right away. There was so much water in the tunnel some of the floorboards had come loose and were floating on top. It would take many days to pump out the water and then they would have to wait for the mud to dry. At least the diggers could rest and catch their second wind. Hasso Herschel took the opportunity to study for his driver's test. Others made improvements in the air or lighting systems. Mimmo Sesta chatted at length with Piers Anderton. Sesta told him that he had no use for governments or leaders of any kind, that people had to take care of problems themselves. "I saw and heard what happened after the Communists closed the border," he said as Anderton took notes. "I saw women in East Berlin weeping because their husbands are in the West, and they will live forever without them. The East German rulers are swine, not because they are Communists, but because they keep people living frightful lives. People should live in happiness, not by an idiotic theory of a future one hundred years from now.

"I must help my friend Peter and his family. Friendship is not just sitting and talking drinking coffee, one must act to help friends and to help anyone whose freedom has been stolen. We must give the East German government no rest, no peace. They should know that there are simple people who want to do something against inhumanity." Sesta's words echoed those of Heinrich Albertz, the West Berlin Senate leader, who had recently compared "courageous" young escape helpers to Nazi resistance fighters.

Anderton took advantage of the lull to film the three organizers reenacting the first two months of the project, before NBC arrived on the scene. They drove around the city in a van, exploring again sites for their project near the Reichstag and Brandenburg Gate. Approaching the Wall at various points, they peered over, then huddled over maps and other documents in an apartment, smoking cigarettes and debating where to start the dig. As always, no voices were recorded.

By now word had spread in the *fluchthelfer* community that

a skilled, hardworking tunnel team had some unexpected free time on their hands. Perhaps they would consider lending those hands and muscles to another dig?

DAN Schorr was disappointed and increasingly anxious, still without a tunnel to chronicle with little more than a month remaining before the Wall's first anniversary. He needed a distraction. Luckily, he found one: Shirley MacLaine. The young American actress, whose career was on the rise after appearing in *Can-Can* and *The Apartment*, had arrived in Berlin for the annual film festival. Attendance was down at this year's event thanks to the absence of cinephiles from East Berlin. There was plenty of star power, however, with the arrival of James Stewart, James Mason, and Maximilian Schell, whose brother had a role in MGM's *Tunnel 28*. Tony Curtis hid out in a secret apartment with new flame (and costar of *Tunnel 28*) Christine Kaufmann.

Schorr, who was single, ingratiated himself with MacLaine, who was married, when he supplied her with pronunciation of a key line in German for one of her festival speeches, a simple *Ich liebe dich*. He ended up as her date for the festival ball, where they were photographed together at a table (she also posed with Jimmy Stewart). Another night, when he picked her up at the Berlin Hilton for dinner, he was amused when a flock of fans surrounded her in the lobby, asking for autographs. This had never happened to *him*. Some of them followed the couple all the way to the restaurant. Schorr asked her if she ever got angry about this. "It will be a lot worse," she replied, "when it stops."

After dinner, the pair motored out to beautiful Lake Wannsee in Schorr's Mercedes. Driving a bit too close to the water, they found the car sinking in wet sand up to its hubcaps. It took them half an hour of walking to hail a taxi. MacLaine had to leave for Rome the next day and invited Schorr to join her. He replied that this was one of the grandest offers he had ever received, to say the least, but that he could not abandon his CBS duties on such short notice. Shirley complained that he was—of all things—too "earthbound."

Back at the festival, Franz Baake, who just weeks earlier had helped connect the Bernauer tunnelers to Schorr's NBC competitor, was awarded a Silver Bear medal for his haunting twenty-eight-minute documentary on the Wall, *Test for the West*.

THE order had come directly from the President, so there was no question in the mind of Secret Service Agent Robert Bouck that it was a serious matter. He was to install a secret recording system in the Oval Office and the Cabinet Room. Three previous presidents had installed listening devices, but they had used them sparingly. Franklin Roosevelt made a few recordings in 1940; Harry Truman and Dwight Eisenhower left behind less than a dozen hours of tapes each. Kennedy's plan would give him far more opportunity than that.

JFK aimed to document face-to-face conversations with aides and visitors, for his own use and/or for the historical record. Without telling anyone why (keeping the *secret* in Secret Service), Bouck ordered high-quality Tandberg tape recorders from the U.S. Army Signal Corps. He stashed them in a file room in the basement of the West Wing and from there ran wires to a pair of microphones in the Oval Office and another pair in the Cabinet Room. At Kennedy's direction, he installed the Oval Office microphones under the President's desk and in a coffee table. Kennedy could activate them with the discreet push of a button on his desk. The microphones in the Cabinet Room were hidden behind drapes and could be turned on and off by a button at the head of the table where JFK sat.

Kennedy had not yet informed anyone about this beyond his private secretary and Secret Service agents. He also had not decided how often to activate the system—which meetings to tape and which to ignore. Would he only record foreign policy debates during a crisis, in case questions arose later? (He had told Agent Bouck that his prime motivation for installing the system was fear of a conflict involving the Soviets.) But what about purely political or campaign discussions? Might they one day reveal him in a bad light? And what would his unaware aides and visitors think if the secret taping system were ever exposed? His brother Bobby had recently joked (with a serious edge) in a note to CIA director John McCone that their father had instructed his sons, "Never write it down." Now JFK had decided, not for the first time, to betray the wishes of Joseph P. Kennedy.

FROM East Berlin came news that Robert Mann, the Stanford student and friend of Joan Glenn arrested back in January, had been convicted and sentenced to twenty-one months in prison. Piers Anderton was among the few American correspondents to attend the brief trial. Mann's attorney, as was typical in such cases, was Wolfgang Vogel. Known for his contacts with West German attorneys, Vogel represented the state and/or the Stasi in court cases and in prisoner exchanges, most notably the "spy swap" earlier that year involving the American U-2 pilot Francis Gary Powers and the Soviet agent Rudolf Abel. Vogel had planned to flee to the West himself in 1953, but the Stasi found out and forced him to become an informer for several years.

Earlier that month, the GDR had staged what the *New York Times* called the first "show trial" since the Wall's construction, clearly sending a signal to *fluchthelfer.* Three men from the West and two from the East had been given stiff prison terms for aiding escapees: five to fifteen years at hard labor. This made Mann's sentence seem a little more bearable. Mann's father flew from California to Berlin and managed to speak with young Robert for twenty minutes, finding him in relatively good health. Afterward he told Joan Glenn that he once questioned why his son had taken such risks to help East German students he didn't even know, but after visiting oppressive East Berlin—he understood.

HARRY Seidel had virtually finished his crude tunnel under Kiefholz Strasse. He was already under the front yard of a house rented by a middle-aged couple named Sendler, who had pur-

portedly agreed to host the escape in the East. Fritz Wagner
had traveled to Belgium to purchase two machine guns for the
mission, but *Dicke* had a problem: they had rescued so many
East Germans via the so-called Pentecost Tunnel the month before that he was now having trouble finding a sufficient number
ready to flee. Wagner's customers tended to be working-class,
and many who had already escaped were finding that jobs were
not easy to come by in the West. Learning this, some in the East
who once wished to exit had second thoughts. Sure, life was no
bowl of cherries in the East, but most did have a job, and certain
basic needs, provided by the state.

Professionals and students were far more optimistic about
making it in the West and also more insistent on personal freedoms. These were the sorts the Girrmann Group specialized in
freeing. Knowing this, Wagner reached out to their office to supply escapees and couriers for the Kiefholz operation. The Girrmann leaders were happy to oblige, working from Joan Glenn's
continually updated list of vetted refugees. The perpetually impatient Harry Seidel didn't feel like waiting a few more weeks
for the breakthrough, however. His mother, still in prison, would
not be able to join this refugee flow anyway and there was a
cycling tourney he wanted to train for. So he quit the Kiefholz
project, while agreeing to offer advice if and when the climax
arrived. Suddenly, with one of his other diggers ill and another
leaving on holiday, Fritz Wagner needed an entire new crew.

Enter LfV intel operative Mertens, who knew of a lengthy
delay at a flooded tunnel somewhere under Bernauer Strasse.
Mertens told Dieter Thieme about the stalled Kiefholz project.
On July 22, Thieme asked his friend Wolf Schroedter to arrange
a meeting with his team. Three days later, at the House of the
Future, Schroedter, the two Italians, Hasso Herschel, Joachim
Rudolph, and Manfred Krebs agreed to finish the Kiefholz tunnel if it passed their inspection. For Herschel and the Italians
there was this extra incentive: they might be able to bring their
loved ones (Hasso's sister) and friends (Peter Schmidt's family)
to the West a couple of months sooner.

On July 27, Schroedter, Rudolph, and Krebs climbed through thick foliage to reach the tunnel's opening. Rudolph and Krebs were shocked to see a familiar figure greeting them. "That's . . . Harry Seidel!" Rudolph exclaimed. Only then did he and Krebs discover what their old middle school friend had been up to for the past year. Seidel was equally surprised to encounter them. All he knew was that "some students" were going to take over his project because their own tunnel had met a leaky pipe.

Harry explained that he had excavated most of this passage but now, "I need a break. You can take it." Handing them flashlights, he led the way into the tunnel. Coming from their own semi-professional dig, they were shocked to find this one had no lighting, no air pipes, no supports. Broken tree roots poked out of the sides in spots. Sandy soil dropped from the ceiling every time a truck drove over Kiefholz Strasse. One could not stoop or even crawl on all fours—diggers and escapees would have to slide on their bellies. The tunnel was so narrow Rudolph worried it might produce claustrophobia in some escapees. (He was feeling a little scared himself.) If there was even a partial collapse there would be no air pocket in which to breathe while you tried to dig out.

It was objectively shoddy work. Harry was brave, and a hard worker, but he was no engineer. This was also his first suburban tunnel, and the longest yet. Now he revealed that fresh air expired at the far end of the tunnel after only a half hour of digging, so you had to shovel in short spurts and then summon a replacement. He also acknowledged the slight smell of gas, perhaps from a leaky pipe—so be careful about lighting matches! Not to mention: No structure, not even a shack, near the opening in which to rest or eat, no toilet, no running water. There was no cart for dirt removal, just those large butchers' pans, *fleischersatten*, with handles attached to ropes. And there were security concerns: dirt from the tunnel had simply been dumped in heaps near the entrance. Bushes and trees seemed to block the view from the East, but who could be sure? To top it all, it was unclear whether the Sendlers had ever confirmed that their house in the East could be used for the breakthrough.

Still, the Bernauer team agreed to work with the Girrmann Group, taking the reins from Seidel. The fact that Harry had started this tunnel mitigated their worries. Harry had repeatedly risked his life fighting the Wall; he was certainly no Stasi snitch. What the tunnelers did not realize was that they were taking a risk just by establishing a temporary link with the Girrmann Group—which *was* infiltrated, at its heart, by Stasi operative Siegfried Uhse.

Within days they had enlarged the cavity significantly. Now an overweight man or woman might actually fit, and children, crawling in the dark with flashlights, might not feel quite so frightened. It would be ready for escape action in just a few days. Maybe Piers Anderton would like to know about it and shoot some footage? It might be a terrific addition to his NBC special—and provide backup material in case the Bernauer tunnel sprung another leak or collapsed. Knowing the tunneling was in good hands, Bodo Köhler and Joan Glenn swung into action, lining up couriers. Their list of would-be refugees now filled several pages.

AFTER weeks of cultivating a relationship with Joan Glenn, Siegfried Uhse's diligence was paying off. Some kind of major escape operation was pending and Glenn inserted him right in the middle of it as a courier. When Glenn summoned him for a walk-and-talk on the street to exchange information, he met her "dressed in my new and chic clothes," as Uhse later told his minder. "I could see from the way she looked and behaved that she found me likable." He could well afford to upgrade his wardrobe, as he was still working as a hairdresser (though no longer at the U.S. base) and as a waiter on weekends, as well as drawing his Stasi stipend.

A week later, on Friday, July 27—just a few hours after the Bernauer team toured Harry Seidel's tunnel—Glenn ordered Uhse to hustle to the East and tell his chief contact among the refugee organizers to "get the group together" for a meeting the following day. And on Sunday evening that person must be ready for a phone call from the West "to receive the essential instructions for the planned escape," set for the following Tuesday near Kiefholz Strasse—although no one was told about the means of escape nor the exact location. Glenn also asked him to find a few quiet spots in the GDR where trucks might park discreetly and pick up about sixty refugees. At this point Uhse did not know about any tunnel. He envisioned armored trucks smuggling people to the West at a remote checkpoint or perhaps smashing through a weak section of the Wall. Glenn, however, stressed that "this escape should take place without the use of violence."

The next day, Uhse's mission was again updated when he visited Glenn at her office (the ban on his House of the Future visits apparently lifted). He was to go East and tell the organizer about a final pre-escape meeting the following evening. Uhse did as instructed, and returned to the Girrmann office. There he noticed a stranger in the next office who spoke in a Swiss accent. When Glenn left her desk for a moment, Uhse overheard the visitor say that the use of trucks in this operation was "not good" because it was "dangerous." Glenn came back and closed the door.

AFTER a weekend in Massachusetts with his family at Hyannis Port, President Kennedy returned to the White House and went directly to the Oval Office. It didn't take him long to inaugurate the secret taping system, activating it for a meeting on a political crisis in Brazil with McGeorge Bundy. Shortly after noon the discussions continued with Bundy, Dean Rusk, and Under Secretary of State George Ball, this time turning to Europe.

"There's nothing else we have to do in Berlin," Kennedy said with an air of resignation. A few minutes later he was complaining about "new recruits" to the diplomatic missions. "I just see an awful lot of fellows who, I think . . . who don't seem to have *cojones*."

Bundy: "Yeah."

That afternoon produced a nearly two-hour discussion on nuclear issues, involving more than a dozen aides and cabinet members. Kennedy had been trying to negotiate a comprehensive test-ban treaty with the Soviets. It would, in theory, slow the arms race by hindering the development of more deadly weapons. The Soviets had balked, objecting to on-site inspections to verify any ban. JFK's fallback position was to at least outlaw nuclear blasts above ground ("atmospheric" tests), in order to reduce the spread of radioactive particles in the jet stream. At the same time, he was pushing a $700 million plan to build or upgrade fallout shelters around the country. Civil defense director Steuart L. Pittman had recently declared that 110 million Americans would probably die following a massive Soviet missile launch, but another 40 to 55 million might be saved if protected by shelters. "Enough persons could live to insure the survival of the United States as a nation," Pittman deadpanned. He admitted, however, that it was difficult for the average American to see how "anything as flimsy as a fallout shelter can stand up to anything as big as a nuclear attack."

Kennedy's latest meeting with advisers ended with several points of contention and very little resolved. A complicating factor: four days earlier the *New York Times* had published an article based on one of the most critical leaks in the President's term so far. It had already led to a nearly unprecedented attempt at media intimidation, carried out at the very hour the White House meeting unfolded. JFK wasn't the only one taping conversations in secret.

THE front-page *Times* article, by renowned military correspondent Hanson Baldwin, had revealed that the Soviets were now "hardening" their missile silos with concrete covers to protect their nuclear weapons in the event of an American attack. Top American officials had kept this from the media and the public— even though the Soviet move offered only limited protection, and the United States was hardening its own silos. Perhaps more than this, what angered the White House and Pentagon was that Baldwin had revealed that U.S. spy cameras high in the sky could detect such retrofitting, citing what he called the new science of "image interpretation" based on infrared and radar images and electronic "emanations."

Such monitoring might reduce the need for on-site inspections, a very good thing, but the Soviets were sensitive to any sort of American spying, and now they would likely try to hide their missile sites from aerial view. Baldwin had also revealed the current number of Atlas, Titan, and Polaris nuclear missiles in the U.S. arsenal, which he claimed gave America a clear edge over the Soviets. Given his long service at the *Times*, Baldwin had credibility where it counted. He had joined the paper back in 1929 and won a Pulitzer for World War II reporting in the Pacific. His views could be described as "hawkish," and he was allowed to express them often in his stories.

The White House quickly moved to find out who was behind the leak, with Attorney General Robert Kennedy ordering FBI director J. Edgar Hoover to take action. On July 27, one day after the Baldwin scoop was published, FBI agents had tapped the phone of a secretary in the Washington office of the *New York Times*. The next day they placed a bug on Baldwin's home phone in Chappaqua, New York. That wasn't all. On the evening of July 30, FBI agents visited the secretary's D.C. apartment and the reporter's Chappaqua residence. Pressed by the agents, the startled secretary told them exactly when Baldwin had conducted interviews for the article, where he had stayed, and the appointments he had made (with more than a dozen Pentagon, military, and CIA officials).

When Baldwin answered his door at the same hour, he told the agents that he was about to sit down for dinner. Asked if they could chat with him after he ate, Baldwin declined, adding that he did not like "this kind of approach." After they left, Baldwin received a call from the most influential columnist in the country, his colleague James "Scotty" Reston—with the FBI listening in, of course.

Reston informed Baldwin about the FBI's visit with the secretary. He called it a witch hunt, an "outrage and we ought to
print the whole thing." Baldwin concurred and said he would
never name his sources. He suggested they contact *Times* publisher Orvil Dryfoos. "This is going very far in this administration—I think it is extremely dangerous," he complained. Reston
said there was a feeling in Congress that "dossiers" were being
kept on certain people. Baldwin agreed this was something new.

The next day, J. Edgar Hoover notified Robert Kennedy that
in light of Baldwin's "resentment and arrogance," no further attempt would be made to interview him. But as for his possible
sources . . . that was another matter. Baldwin met with Dryfoos
(with Reston on speakerphone) to consider exposing the administration's assault on the free press. They decided as a first step
that Reston should call McGeorge Bundy and press for details
on the scope of the FBI inquiry and who had initiated it. That
evening, Robert McNamara stopped by Reston's home for a
friendly two-hour chat. He apologized for the rough tactics but
also called the leak a "clear violation of the law." He did not
identify which law.

ON the final day of July 1962, Daniel Schorr finally got his tunnel. His big break did not come via the Girrmann Group. His
savior was James P. O'Donnell.

Boston-born Jim O'Donnell was an old Berlin hand. He had
even met Hitler as a student in Germany in the 1930s. After he
graduated from Harvard and served in Europe during World
War II, *Newsweek* sent him to explore the bunker where Hitler died, where (after bribing a Soviet soldier) he found files,
notebooks, and diaries missed by other searchers. He covered
the Nuremberg trials and the Berlin airlift as the magazine's
bureau chief. For most of the 1950s he served as an editor and a
writer, again often on Germany, for the *Saturday Evening Post*.
A longtime friend of the Kennedys, O'Donnell worked on JFK's
1960 campaign and sent the candidate advice on Berlin. The

following year he was asked to head a new "psychological warfare group" at the State Department that would feed anti-Castro propaganda articles to Latin American media.

When the Berlin Wall first appeared, O'Donnell called for the United States to tear down the barbed wire, declaring, "This is bigger than the Bay of Pigs!" The idea went nowhere. The White House decided instead to send Vice President Lyndon Johnson to Berlin to signal American resolve. O'Donnell played a key role in persuading Kennedy to also invite retired general Lucius Clay on that trip. (LBJ had complained, "There'll be a lot of shooting, and I'll be in the middle of it. Why me?") When the President later appointed Clay to oversee U.S. operations in Berlin, O'Donnell served as a top aide to the general. With Clay summoned home to America in the spring of 1962, O'Donnell returned to writing, while remaining one of Berlin's most prominent American fixers and men-about-town. Despite his quick temper and bouts of logorrhea, he remained a key source for the U.S. Mission on what German journalists and average citizens were thinking.

Now, in late July, he was alerted to a new escape project by Rainer Hildebrandt. He was a friend of Harry Seidel, and a well-known anti-Communist agitator. Hildebrandt told O'Donnell that a certain Fritz Wagner was looking to raise funds for a tunnel by selling film rights. O'Donnell knew all of the veteran print and TV journalists in the city, but he approached CBS's Schorr first—perhaps because Dan had begged him to do this if he ever heard about such a project. O'Donnell asked whether Schorr would like to attend a meeting the next day to discuss the terms of a deal with himself, Wagner, and Hildebrandt.

Schorr, his months of fruitless inquiry at an end, readily agreed. And if the tunnel broke through before August 13, he might still get his bombshell special report for the first anniversary of the Wall.

7

Schorr and the
Secretary

AUGUST 1–6, 1962

The meeting between Daniel Schorr and Fritz Wagner, brokered by James O'Donnell, to arrange the sale of TV rights for the Kiefholz tunnel proceeded as planned on the first day of August. Notably absent were the tunnel's creator, Harry Seidel, and the Bernauer crew, currently risking their lives without pay to complete his work. Wagner claimed that at least forty-five refugees were scheduled to crawl to freedom in a few days. He demanded 100,000 DM ($25,000) for the rights to film the final preparations and escape. Schorr hooted at this, offering 5000 DM instead. At a second meeting that day, he sweetened the deal after Wagner told him that one of the refugees would bring exclusive footage shot on August 13 the previous year, perfect for the newsman's scenario for a first-anniversary special. Then Schorr left to consult his German cameraman.

If U.S. diplomats were still in the dark about a pending mass escape through a tunnel at Keifholz Strasse, this ended on August 3. That morning Jim O'Donnell arrived at the massive L-shaped structure that housed the U.S. Mission and the offices for the American military's Berlin Brigade. The complex on Clay Allee (named for General Lucius Clay, hero of the Berlin airlift) had previously served as headquarters for Hitler's Luftwaffe. Although the Nazi insignias had been removed from its façade, some of the ornamental cement eagles remained.

O'Donnell asked staffer Ralph A. Brown for the Mission's help in obtaining information on two Germans he had met a few days earlier, Rainer Hildebrandt and the man he called "Warner" (obviously Wagner). He described the escape plan and the negotiations that led to Dan Schorr handing over more than 5000 DM. O'Donnell held little back, even disclosing the tunnel's location as far as he knew it. The diggers, he said, were just a few yards from their breakthrough point, with the escape scheduled for Sunday, August 5. And Schorr had already sent his cameraman to the tunnel at least once, the day before.

Why was O'Donnell, a journalist, revealing all this to a Mission staffer? He explained that his principal motive was "to see to it that proper publicity is given to an escape attempt." Whether this would come in the form of an exclusive article written by O'Donnell himself, Brown could not guess. Perhaps O'Donnell was merely testing the Mission's receptivity to any sensational mass escape: for or against? Or he naïvely thought the diplomats would agree to promote it. Brown, deeply troubled by the visit, sent a memo up the food chain at the Mission. It concluded:

I expressed concern about these plans and recommended to O'Donnell that he drop his contacts with the group. I pointed out that the American press had previously been accused of exploiting escapes from the East and that this had an adverse effect on our relations with the German press and Berlin city [officials]. I told him that this tunnel operation has probably been blown to the Eastern authorities and could probably not be carried out successfully.

Oddly, Brown did not reveal *why* he thought the tunnel escape had "probably been blown." Who or what was the source for this opinion? (Even Siegfried Uhse at this point possessed few details on the operation.) Brown did enclose something else with the memo: a rough sketch, on lined paper, probably drawn by O'Donnell, of the area where the tunnel was located. It traced the path from a hole in the ground in the West, under the Wall, to an unidentified structure just across the border—all of this below the S-Bahn tracks and not far from the station at Baum-schulenweg, which was, oddly, a stop or two from the actual es-cape site. Was O'Donnell misinformed—or was he deliberately passing false information to the Mission (and if so, why)?

The Brown memo got the attention of Arthur Day, who ran the Mission's "politics" desk. He dashed off a cable to a colleague at the U.S. embassy in Bonn, including much of O'Donnell's information. "Schorr reportedly plans filming escape for use in documentary on August 13," Day warned. He advised:

> We consider it would be most damaging to US interest to have association Schorr and O'Donnell become public. Par-ticularly given plans for documentary appears highly likely will become public. We intend attempt persuade Schorr and O'Donnell disassociate themselves and not use films of escape in any way that would indicate any association. If Schorr does not agree, we may recommend CBS in US be ap-proached by department. In view of inability talk to you on telephone, and urgency of situation, we proceeding arrange meeting with Schorr and O'Donnell this afternoon.

His colleague replied that he had discussed the situation with two of his superiors, including Ambassador Walter Dowling. "They naturally approve your attempt to dissuade O'Donnell and Schorr from association with escape," he advised.

Brown, after getting this go-ahead, passed the problem on to Charles Hulick, the Mission's deputy chief. He contacted Schorr swiftly, but the cantankerous CBS correspondent remained de-termined to film the escape. After a routine press briefing the following afternoon, Hulick broached the subject with Schorr again. Hulick impressed on Schorr the risks that the "other side" knew about the operation and, if CBS went ahead with the filming, that the Communists might seize his "documentary evidence" and even some of "his" people. Yet despite repeated (though unsourced) claims that the tunnel had been compro-mised, there was still no indication that anyone in the Berlin Mission had warned the diggers or Girrmann staffers about these risks. Perhaps they didn't really believe the mission was doomed, or didn't particularly care if it was. What they were certain of was: they didn't want any TV coverage of it, let alone a prime-time documentary. Schorr replied that he meant what he told Hulick the night before—he would at least weigh the matter.

Shortly after this, Jim O'Donnell informed the Mission that the escape action was now set for August 7 at 6:00 p.m., and that the number of likely refugees had swelled to about ninety, easily the biggest one-day tunnel exodus ever. He said that Schorr had decided to return to Bonn, but that his cameraman in Berlin might be at the scene for the breakthrough. It appeared that Schorr had not scrubbed his coverage, just taken himself out of harm's way.

In fact, Schorr still planned to be in Berlin for escape day.

Hulick cabled a report not only to the Bonn embassy ("in case embassy wishes approach Schorr in Bonn tomorrow") but also to the head of the Berlin Task Force in Washington, which coordinated policy for the President. This signaled a new level of concern. Hulick tempered the claim that Schorr might be pulling back by noting O'Donnell's "uncertain reliability." Perhaps the President should be advised?

OUT of the loop about all this, the diggers at Kiefholz Strasse continued widening Harry Seidel's tunnel, despite low oxygen in parts of the passage. Unsure if Harry had accurately pointed the tunnel at its target, one of them crawled to the end and pushed a metal bar through the tunnel's ceiling to a foot or so above the surface. A colleague with binoculars watched for its appearance from a low hill in the West. When the metal rod appeared, the diggers knew they were in the Sendlers' front yard, only a little off line. Fortunately they had time to shift the tunnel a bit to the left. Soon they would be poised for a breakthrough. Would the couriers and refugees be ready?

Over at the House of the Future, an evening meeting convened to discuss the escape operation. Attending were activist leaders Bodo Köhler and Detlef Girrmann; coordinator of couriers Joan Glenn; Hartmut Stachowitz, a veterinarian who wanted to get his wife and son out of the East; Rudi Thurow, a former East German border guard, now an active Girrmann associate; a twenty-year-old East German woman who had just been smuggled to the West in, or under, an automobile; and a female student from Switzerland who had just returned from the East after notifying some folks about the upcoming operation. Just another day at the office.

As the meeting proceeded, Köhler took a few phone calls. After one, he informed the group that he thought "somebody" from outside had monitored that conversation. Probably he heard a clicking sound or heavy breathing. When Stachowitz asked what he meant, Köhler suggested that an American "department" in the West was intercepting their calls.

PRESIDENT Kennedy switched on his own secret listening device as he sat down with his Foreign Intelligence Advisory Board in the Oval Office. Present were Dr. James Killian (former president of MIT), prominent attorney Clark Clifford, General Maxwell Taylor, Robert Kennedy, and Dr. Edwin H. Land of Polaroid fame (who helped design the U-2 spy plane and its cameras). The subject: that recent leak in the *New York Times*.

JFK had approved the FBI probe and taps on the phones of Hanson Baldwin and a *Times* secretary, but also convened this high-level panel. Killian, the chairman, quickly declared it "the judgment of your board today that this is one of the most damaging unauthorized disclosures and leaks ... a tragically serious breach of security." The discussion quickly shifted, however, from catching the culprit to policing the bureaucracy. Killian called for "drastic and unprecedented procedures" in light of the fact that leakers "have no real fear that any punitive action will be taken." He proposed a new policy for the Department of Defense and other agencies that handled classified information:

after any interaction with the media, a staffer must file a memo with his superior revealing whom he talked to and when, and what was discussed. As Killian put it, this would make the potential leaker "more careful, more reticent" and "leave the man in a much more vulnerable position to that punitive action in the event that it is necessary."

"So the burden would be on the government rather than on the newspaperman," Kennedy affirmed.

Killian claimed, improbably, "We don't see that the press could have an adverse reaction to this, at all. Because this is none of their business. . . . You are not interfering with their access to personnel." Moreover, he suggested, a new office should be devoted to such oversight: "an expert group that would be available at all times to follow up on security leaks." Its very existence would create a "deterrent effect."

Clark Clifford, JFK's former lawyer, backed the idea with enthusiasm, and urged that they go beyond investigating after the fact. They should follow the *pattern* of leaks, to the point of tracking the activities of certain journalists. "It can be done quietly, unobtrusively," he claimed. Perhaps this level of press monitoring hadn't been required in the past, but leaks were more damaging now. "There are many things," he added, "that such a sensitized group could do that—they could follow the press and see evidence of—"

The President interrupted: "That's a very good idea. We'll do that."

Clifford continued, "Let's then begin to get up a file on these different men [reporters]. . . . To my knowledge it's never been done before and it is long overdue."

That was all right with the President. The press "are the most privileged group. . . . They regard any action in this area as a limitation on their civil rights," he said. "And they are not very used to it."

JFK asked for a draft of a protest letter, over his name, to be hand-delivered that week to *New York Times* publisher Dryfoos, written in such a way that it would "demonstrate that this is not an overly sensitive administration." Everyone laughed, as he probably intended, and the meeting broke up.

OVER the next few days, the White House began receiving urgent messages from the State Department on a very different media crisis: sensational coverage of a new tunnel under the Berlin Wall was expected from one of the President's least favorite "privileged" journalists, Daniel Schorr. Cables from Berlin and Bonn outlined the potential for an international incident. The tunnel project might be exposed by both East Germany and CBS News. It was the kind of sensitive issue Dean Rusk always discussed with the President. Rusk's press managers knew that Pierre Salinger would also want to hear about this matter directly. Salinger, as JFK's proxy, insisted on signing off on any moves against the media. He had extraordinary access to Kennedy, having served as his press secretary since the 1960 campaign. One way or another—from Kennedy or Salinger— approval of a shutdown move against Schorr was almost certainly at hand.

Rusk, like his boss, was no fan of the press. Too often reporters undermined U.S. policy simply to make waves. In that sense, they were not patriots, and they refused to consider the big picture. Yes, the public had the "right to know," but in Rusk's view this must be limited by the right of officials to conduct business responsibly, in the dark. He could never understand why, if a reporter extracted a secret from the State Department and gave it to the Russians, he would be rightly labeled a traitor, but if he disclosed it in a newspaper that journalist might win a Pulitzer. For all of these reasons Rusk had learned to down a shot or two of scotch to relax before his press conferences. Then he might actually enjoy fencing with reporters, rather than simply invoking one of his favorite quips, "There are times when a secretary of state must say nothing—at considerable length."

Now he was agitated enough to ask James L. Greenfield, his number two in State's public affairs office, to meet with Blair

Clark, the CBS news director. Rusk briefed Greenfield about Schorr's involvement with a tunnel in Berlin, stressing that the mission was almost certain to fail (he didn't say how he knew this) in a fashion that would be "too public" and "raise tensions." Lives were "endangered." As an old friend of Jack Kennedy, Clark would likely give them a fair listen. They also knew that Schorr was not in JFK's good graces, and that the President abhorred media adventurism that threatened national security.

Greenfield, once one of *Time*'s leading foreign correspondents, considered Rusk extraordinarily sensitive about confrontations with the Soviets in Berlin, so this move made sense. But while Greenfield was friendly with Blair Clark (a fellow Harvard grad), he knew it was highly unusual for a TV news director to welcome a visit from a State Department press officer. Still, in his view, the meeting to discuss Schorr's project went very well. He told Clark lives were "being compromised." With just hours remaining before the escape operation began, Clark said he would seriously consider ordering Schorr to abandon his tunnel.

AND there was more trouble in store for Schorr. Piers Anderton had been tipped off to his plan to film an escape near Kiefholz Strasse after hearing that other journalists had found film packaging near a tunnel entrance in the West—a particular type that CBS supplied its cameramen. What to do? Anderton broke Reuven Frank's strict non-contact rule and tried to reach the producer, but he was visiting family in Montreal. NBC news chief Bill McAndrew, deeply concerned about getting scooped by the network's chief rival, did manage to reach Frank. He asked him to consider ending his vacation and flying to Berlin to learn more about what he called "the CBS Tunnel." Frank decided to wait it out.

Now Anderton learned from one of the Bernauer tunnelers the time and place for the escape, even the location of the target house on Kiefholz Strasse. This was a godsend. Anderton had been involved with the tunnel at Bernauer Strasse for more than two months, with its fate unknown, particularly after the great flood. Even if the rest of that project went well, CBS seemed poised on this August day to take the edge off his program. But what if NBC could negate its rival's advantage—by scooping CBS's scoop? From Bonn, Anderton hastily arranged film coverage across the border from the Sendler cottage, from a window high up an abandoned railroad-switch tower. In his appointment book under the date August 7 he wrote, simply, *Tunnel*. Then he caught a late plane for Berlin.

IF Free University student Wolf-Dieter Sternheimer wanted to get his fiancée, Renate, out of the East—and he did, very much—he would have to perform a few key chores for the Girrmann Group. As the date for the operation was pushed back to August 7, Detlef Girrmann sent Sternheimer to the East three times to update a certain shop owner who wanted to flee. Girrmann also asked him to find a truck and driver in the East to haul the ever-expanding number of escapees.

Then it got complicated. Renate told her friend Britta Bayer about the escape plan, and she asked *her* fiancé, Manfred Meier, to look up Sternheimer in West Berlin. Sternheimer said that Britta could join the flight of refugees if Meier arranged one of the trucks to be used that day. Meier promptly found an East Berliner who promised to borrow a truck for this purpose—if *he* could also take the tunnel to the West!

Meier was no stranger to subterfuge, having helped a group of nuns free about a dozen East Germans (the sisters swapped clothes with the refugees) the previous year. Now Girrmann asked him to scout the Kiefholz area, trusting this stranger with the address of the Sendler cottage, which was behind "two big gates." Girrmann gave no order to visit the Sendlers to confirm their consent. Meier thought this was all very odd, even disturbing. Why was Girrmann—the man in charge of organizing the refugees—so ignorant about the site on escape eve? He was asking Meier to identify the location of such basics as the entrance gates, the backyard, and the doors to the house.

In any event, Meier made it safely to the Sendler property and back to the West. At the House of the Future he helped sketch a map displaying the paths and gates at the escape site, and spots along nearby streets where trucks could drop off passengers. Girrmann kept asking him to locate a certain tree, perhaps as a signpost for the refugees. Meier thought, *A tree? Really. You have dozens of people coming to this site and no one until now bothered to check it out and now you're asking about a tree? Are you crackers?* It all seemed surprisingly amateurish. He began to worry about what might happen to his fiancée, Britta, the following day.

AFTER August 6, Siegfried Uhse could be forgiven if he changed his Stasi code name from "Hardy" to "Lucky." It began when he called Bodo Köhler at his office that afternoon, just checking in. Köhler had been suspicious about Uhse for a long time. Just the week before, Joan Glenn had asked Uhse to find some "discreet" spots where refugees might safely meet trucks that would carry them to an escape point. He proposed a couple of sites—fed to him by his Stasi minders—but Köhler wondered why, under questioning, Uhse didn't seem to know details about them. Glenn reassured him that Uhse was trustworthy. Now he asked Uhse to come to the House of the Future as quickly as possible.

When Uhse arrived, Köhler told him that the escape was set for the following day. Girrmann claimed that the occupants of the target house had agreed to the action. Still, given the daring plan and large number of refugees, this might be, Girrmann admitted, something of a suicide mission. Köhler asked Uhse to take a taxi to visit the courier Sternheimer, who would tell him what to do. Time was short and couriers few, so any worries about Uhse's true motives receded.

Sternheimer instructed Uhse to visit a woman who was helping to arrange a truck in East Berlin. Uhse was to ask her these questions and then report back to the House of the Future: Will the driver and his truck be ready for the escape the following afternoon? Did the driver get the tarpaulin to cover the back of the truck? Where should the escapees meet the truck and what are the passwords? Has the capacity of the truck changed and is it possible to squeeze in more refugees?

Uhse now knew for certain that the escape action was scheduled for the next day. He didn't know exactly where the refugees would meet the trucks, nor where the trucks would then go, but he was suddenly at the very center of the operation—and set to meet Sternheimer again first thing the following morning. After hurrying through a checkpoint to the East to meet the woman, he stopped to visit his handlers. The Stasi learned, at nearly the last minute, that they might be able to destroy the most ambitious Berlin escape action yet.

WORD was just starting to spread to potential refugees about the escape the next day. One of the first to learn about it was Hasso Herschel's kid sister, Anita Moeller, who lived in Dresden with their mother and would need to trek to East Berlin for the big event. Hasso had told her he would alert her via a coded telegram—along the lines of *You can pick up your tickets for the opera tomorrow at 3 p.m.* On August 6 the wire arrived. It did not reveal what the method of escape would be, which was probably a good thing since Anita considered herself rather claustrophobic.

When Anita had begged her brother to take her with him to the West less than a year earlier, he had insisted that she stay with their mother, promising, "I will get you later." Anita was working part-time for her architect brother-in-law, but her urge for going had only increased. A former member of the Young Pioneers, the primary Communist youth group, she had long since come to realize that she had swallowed nothing but lies. She wanted the kind of life she had seen on visits to her mother's family in Cologne in the West. Now, when she saw photos of soldiers at the Brandenburg Gate with weapons pointed east, not west, she took it as a symbol. The Wall was a Maginot Line, in reverse.

She would be fleeing with her husband, Hans-Georg, and sixteen-month-old daughter, Astrid. Her marriage was faltering; the couple had wed only after Anita became pregnant and had lived apart much of the time since. When the coded telegram from Hasso arrived, the couple packed a few things and left for Berlin, where they would stay with friends who also planned to escape. There, so close to a reunion with her brother, nerves and excitement took over, and a night involving a little too much wine began.

ON the evening of August 6, after a sultry 90-degree day in D.C., Dean Rusk was so concerned about a possible tunnel fiasco in Berlin that he stayed very late in his seventh-floor office. At ten minutes after ten he talked on the phone with Charles Hulick at the Berlin Mission, seeking an update on the escape and, more important, on Daniel Schorr's plans to cover it, despite multiple warnings from Rusk's underlings.

A few minutes later, Rusk conferred with the President's press secretary, Pierre Salinger, ensuring that Kennedy was on board with this intervention. At 10:50 p.m., Rusk met with the State Department's top two men in public affairs, Robert J. Manning and James Greenfield. Three men from the CIA joined them. Finally at 11:25, Blair Clark, the CBS news director, arrived, at Greenfield's invitation.

Born in 1917, Ledyard Blair Clark grew up in Princeton and met John F. Kennedy when they lived in the same house at Harvard. Clark edited the *Harvard Crimson*, then worked for a St. Louis newspaper and helped JFK revise part of his book *Why England Slept*. When he got out of the army, he bought a newspaper in New Hampshire whose star reporter was Ben Bradlee. After Clark sold that paper, he joined CBS News in 1953 as a Paris correspondent, producer, and radio anchor. He remained a Kennedy pal, even joining young Jack and Frank Sinatra on one occasion when they partied in Las Vegas.

After JFK was elected president he asked Clark to serve as a consultant on how best to exploit the medium of television. This would compromise Clark's position as an allegedly nonpartisan network newsman, but he accepted. Kennedy also offered Clark the post of ambassador to Mexico, but he turned it down to remain at CBS, where he was elevated—some said because of his White House connections—to vice president and general manager of the news division. He quickly made his mark, hiring a number of promising young correspondents (including Dan Rather), but it was Clark's friendship with JFK that won CBS the chance to produce one of the highest-rated specials ever aired on television: Jacqueline Kennedy's guided tour of the White House in February 1962. It was another two-way favor between a TV network and America's telegenic president.

Now Dean Rusk was asking Clark to order Dan Schorr off the tunnel story. The State Department would even provide a secure phone line to make the call to Berlin. Salinger and Rusk—and the CIA experts—spoke of "proof" that the tunnel had been compromised and that lives were endangered. With time short and no way to evaluate these claims, Clark agreed to talk to Schorr that night.

WHEN Schorr got a call at dawn in his Berlin hotel room, he was perplexed. He was being summoned to the U.S. Mission. On arrival, greeted by a U.S. Marine guard, he was even more startled to learn that he would be speaking on a secure line (indicating some sort of top secret issue), arranged by the U.S. military. It wasn't so surprising that the man on the other end was his boss, Blair Clark. But what was he doing in Dean Rusk's office around midnight in Washington?

"What's this I hear about you planning to film a Berlin tunnel escape?" Clark asked.

"I told our foreign editor all about it," Schorr replied.

"Well, I am sitting here with the secretary of state in his office," Clark said.

"Yes?"

"And he has convinced me that you shouldn't go ahead and do that."

"Why not?"

"Because it would be considered a provocation, it could lead to a great deal of trouble, and the State Department doesn't want any unnecessary trouble at the Wall."

"That's the trouble with the State Department. That's why there *is* a wall."

"Dan, I know you don't like to be ordered around," Clark said, putting it mildly, "but that's it—I want you to scrap all your plans to do that film."

Schorr was stunned. "Okay," he said, "but would it make any difference for you to know that once this gets around and we don't do it, they [the tunnelers] will go to NBC or, god forbid, ABC?"

"It's an order."

The call lasted just six minutes. Schorr returned to his hotel humiliated and fuming. The whole concept was wrong—this administration, any administration, dictating news coverage. He knew Blair Clark was a Kennedy man, and he figured that JFK had talked to him, possibly even pressured him, which made Dan even angrier. But there was nothing he could do about it now. Neither Schorr, nor the White House, knew that another American news team would soon be in place to film the escape.

Dean Rusk finally got a chance to go home past midnight, but not before dictating a cable, marked "eyes only," that was sent to Hulick at the Berlin Mission shortly after one o'clock.

I saw Clark tonight and he agreed scrub CBS participation in tunnel project. However there is evidence that this matter has been the subject of communication between Schorr and his home office over period of time. I am disturbed that so many people aware of project. I cannot assess matter fully here and must leave to local discretion. But urgent consideration should be given to steps to alert East Germans involved to high probability that secrecy broken and they

walking into trap. If possible should be considered whether
not advisable scrub whole effort.

You should also consider quiet surveillance of area to ensure that photographers and others not poised for tunnel exit.

A copy of the cable would be sent to Mac Bundy and to Pierre
Salinger, who would brief the President on its contents. As usual,
no reply from the White House would signal approval.

8
Kiefholz Strasse

AUGUST 7, 1962

Siegfried Uhse had evolved from marginal to central player in
the Girrmann Group's latest rescue operation. Here he was, in
West Berlin at 7:30 in the morning on escape day, meeting fellow courier Wolf-Dieter Sternheimer at Moritz Platz to learn the
address for the final pre-mission meeting. Sternheimer had been
awake since 4:30, when Detlef Girrmann and Dieter Thieme
showed up at his apartment to deliver the map of the area around
the Sendler house created the day before. At the top of the sheet
Sternheimer jotted down the signals Girrmann associates on the
street would flash to refugees as they arrived on trucks or on
foot, and he was ordered to share it with couriers in the East.

At his early morning meeting with Uhse, however, Sternheimer decided to hand him the crucial map, explaining that he
didn't want to get caught with it at the Heinrich-Heine checkpoint later that day. Uhse, of course, was the worst person in the
world he could have entrusted with this gift. The map sketched
the route the trucks would take as they neared their final destination that afternoon, along with street names, drop-off points,
the location of two gates in the rear of the Sendler property, and
a square marking the tunnel breakthrough point. "Orient yourself a little," Sternheimer suggested.

Minutes later, Uhse, who had decided to chronicle this day

in a few handwritten notes, wrote: *Girrmann does not know that I got the sketch from Sternheimer.*

At the meeting for Girrmann couriers and organizers, the slim, shy Uhse, dressed in a dark nylon jacket, sat largely unnoticed. He learned that the escape would occur between four and seven that evening. Three trucks would bring East Germans to the tunnel site. One would take on passengers outside a cemetery in the Lichtenberg area and another near a school at Weissenseer Weg. The site for the third truck, under the direction of Mimmo Sesta, had not yet been determined. Refugees were to show up at an appointed time and identify their truck by a piece of white tape on the passenger's side of the windshield. They would ask the driver about a nonexistent street. He would reply, "The street must be here close by." Then they were to ask for a lift.

When fully loaded, the trucks would follow a specific route through a series of Girrmann checkpoints near the escape site. Other refugees walking to the area would reach the same spots. Girrmann reps on the street would signal them: *Combing hair*—continue driving. *Blowing nose*—come back in ten minutes. *Tying shoelace*—danger, leave immediately. If the escape was still on, the refugees, in trucks or on foot, would pass through the two gates and end up in the rear of the Sendler house, which was almost directly on the border. Someone would lead them to the house (no address given for security reasons), where tunnelers would guide them on their way West.

Before the briefing broke up, Uhse was assigned the task of meeting two of the couriers—Sternheimer again, and a man named Stachowitz—that afternoon to go over the signals and to find out where the third truck would be waiting, which he would then report to the House of the Future. Uhse updated his notes. One of the attendees had been a short Italian (clearly Sesta), "age ca. 24" who had curly hair and was doing "a lot" with the trucks. A second was a "short, fat man," seemingly "influential," who was called *Dicke.* He was supposed to arrange a truck and driver for the Italian. Then there was the Swiss woman he had seen at the House of the Future many times before, and three young men (two of them "wearing studded blue jeans").

Notes in hand, Uhse rushed off to meet his Stasi bosses. Thanks to the map, they would know where the breakthrough was to transpire, along with all the checkpoint signals. Of the escape site, Uhse told his handlers: "The house is said to be inhabited." Were its residents in on the plot? Still unclear. "During the meeting," he reported, "it was mentioned that their connection to the house has been broken and that there is no knowledge as to whether the wall had been broken through yet and if everything is going to be in order."

ESCAPE day dawned overcast, a little cool for August, with the threat of rain, and it had not started well for Hasso Herschel's sister and family. Anita Moeller and her husband had overslept after sharing too much wine with their friends, and now worried that they would miss their appointment with the truck taking them to the border crossing. Their hosts somehow knew the location of the Girrmann checkpoints near the escape house, so they decided to take their time and go there on foot. It seemed a hell of a lot safer than hiding in a truck anyway. As Hasso suggested, Anita gave her daughter part of a sleeping pill so she would stay quiet during the upcoming adventure. It worked for a while, but then the toddler woke up and, crawling on the floor, kept nodding off and hitting her head.

Peter Schmidt, his wife, Eveline, and their infant daughter were also on the escape roster, thanks to the Bernauer tunnel crew. (Their exit would require a wrenching break with Eveline's grandparents, who had raised her after her mother fell ill with tuberculosis.) The Schmidts had received word a few days before to stay close to home because an escape plan was nearing fruition. But when the courier arrived on the afternoon of August 7, Eveline was running errands. When she returned from the tailor shop, Peter berated her for making them late. "We have to go!" he shouted. "We have to go now! There's a tunnel! It's happening today!"

After frantically gathering clothing, IDs, a change of diapers, and their child, they bid farewell to the small house and property. There was no time to reach the rendezvous point for the truck, so they set out for a one-hour trip on the S-Bahn, followed by a twenty-minute walk to the address they had been given for the first Girrmann checkpoint, on Puder Strasse.

Elsewhere in East Berlin, near the city's largest lake, the Müggelsee, another toddler was getting prepped for the flight to the West. His parents, Hartmut and Gerda Stachowitz, had lived separately off and on for years to expedite their university studies—Hartmut at the Free University and Gerda in Dresden—but they remained devoted to each other. Now, trapped behind the Wall, she could no longer see Hartmut in the West, although he visited her occasionally thanks to his West German passport. He had assumed Gerda would be allowed to join him, but every petition they had filed had failed, including an appeal to the German Red Cross, to bishops in the church, and through her father, an award-winning director of a GDR fisheries institute. (The Stasi would respond by investigating the father and the institute.)

Early in 1962, Hartmut got Gerda's name added to Joan Glenn's list of potential refugees, and in July he volunteered to act as a courier on escape day. Yes, exiting by tunnel was dangerous, but this might be his wife's and son's last chance. He just had to meet with Sternheimer on the afternoon of August 7 to learn the signals, the location of the trucks, and his courier assignment.

When Stachowitz arrived near an East Berlin S-Bahn station at one o'clock, he discovered that the quiet young man in the nylon jacket he had seen at the House of the Future the night before was also present, but didn't know his name (which was, of course, Uhse). Sternheimer gave Stachowitz the location of one of the transport trucks and the address of a married couple to notify in the East, then left to make sure the truck driver was ready. The operation was set to begin about 4:30. Stachowitz took the S-Bahn far to the East. Finding no trace of the couple, he moved on.

Locating his wife at her parents' house, Hartmut helped her pack and get ready to hustle to the escape truck. He would accompany Gerda and their child to the truck to make sure all was well. His only instructions to her: *If you see any police or border guards during the escape, go home immediately. If they draw weapons, raise your hands and surrender.* Then they loaded their child in a stroller and told Gerda's parents they were meeting some friends for coffee.

MIMMO Sesta had decided to address the desperate need for a third escort truck himself. Fritz Wagner had directed him to a butcher at an East Berlin market. Mimmo crossed the border, but the man was nowhere to be found. (The way Siegfried Uhse heard it: the truck showed up but the driver, perhaps recognizing the risky nature of his mission, drove off.) Obtaining a truck in East Berlin was extremely difficult, especially one with a tarp to cover human cargo, and failing to find one was not an option on this particular day. So Sesta prowled the market seeking any old friends of Wagner. Fortunately, he came upon someone who knew *Dicke*—and had access to a truck. Sesta discovered that the man wanted to exit the East, so he said, *Well, guess what, you can drive your truck to freedom this very day!*

Back in the West, as the day stretched into midafternoon, Manfred Meier was growing nervous. He had expected one or more of the couriers to have returned to the Girrmann office by now with word that the operation was going smoothly, but that had not happened. With his fiancée, Britta, set to walk to the meeting point near the Sendler property any minute, he decided to rush to the East and see if refugees had arrived at the site—and whether police or Stasi were waiting for them.

THANKS to Siegfried Uhse's report and the map from Sternheimer, various members of the GDR defense force—police, military, Stasi—swung into action. They knew approximately where the escape house was located but were not sure of the exact site.

At twenty minutes past three, the commander of the military's First Border Brigade ordered a squad of soldiers, an armored personnel carrier, and a truck with a water cannon to a staging area in Treptower Park, just blocks from the Sendler house, where they were to "wait, staying under cover," as the brigade's log put it. Two other captains were ordered to "initiate all measures for the prevention of a border breakthrough." This included, according to the log, "heightened alertness" along the border, "intensified monitoring of the enemy's territory" near the construction site where the tunnel most likely began, and deployment of additional soldiers. Four American jeeps were observed across the border, with three machine-gun-bearing soldiers in each. A black Mercedes (belonging to Fritz Wagner) soon arrived as well.

The MfS was also mobilizing, in coordination with the military. They were to, as the Stasi log explained, "prevent a presumably planned border breakthrough near Kiefholz and corner Puder Strasse." Around four o'clock, two Stasi comrades approached what they considered the likely arrival point for the refugees, the Sendlers' backyard. Spotting them, Friedrich Sendler went out and locked the wooden gates. The Stasi men backed off, but an hour later a commander ordered agents to return and force the issue.

At the same time, two blocks north on Puder Strasse just across from Treptower Park, three undercover Stasi operatives were mingling with five civilians near the spot where Uhse's map had indicated trucks were to drop off refugees for the short walk to the narrow lane leading to the rear of the Sendler property. When Manfred Meier arrived at the tram stop near Puder Strasse, he spotted what he took to be plainclothes Stasi agents checking the IDs of anyone entering or exiting the side streets. Someone, he realized, must have tipped them off.

Pacing up and down the street, he kept an eye out for his girlfriend. When he spotted her, arriving on foot along Puder Strasse and accompanied by Sternheimer's fiancée, he strode toward the two on the sidewalk. As he passed, without slowing, he whispered, "The operation is busted, leave immediately." The two women pivoted and exited the street. Meier felt relieved, but feared he had become conspicuous—and knew that the Stasi would question what any West German was doing on this block at this moment.

Sure enough, he was detained when he tried to leave the street. The Stasi confiscated his West German passport, asking, "What are you doing here?"

Meier protested, explaining that he was in the area to visit a cousin. "Let me talk to your superior!" he demanded, to no avail. In saving his girlfriend he had sacrificed himself, earning the distinction of being the Stasi's first arrest at the escape site that afternoon.

At 5:15 p.m., two other West Germans approached the undercover operatives and, apparently mistaking them for Girrmann Group associates, asked whether the operation was still on, as they had not seen anyone at their truck pickup location. They had two refugees in their car—would these escape helpers (the Stasi agents) take charge of them? All too willingly. Rather than blow their cover at this point, the Stasi let one of the West Germans leave while engaging the other in conversation. The remaining man promptly confirmed that the site of the tunnel was inside the house of Sendler, the carpenter, over on Kiefholz Strasse.

JUST across the border, in the late afternoon shadows beyond a tangle of bushes and trees, the three-man breakthrough team reported for duty. Hasso Herschel, like Harry Seidel an athlete and a daredevil, volunteered to lead the crew and be first through the hole into the Sendler cottage. Understandably, no one contested this. Hasso had not done any digging on the tunnel, but he was motivated by the belief that his sister was heading to the scene at that very hour. Joachim Rudolph and Uli Pfeifer did not have friends or family among the potential escapees, but they accepted Herschel's request to assist him. Hasso had known Uli for years and he recognized Joachim as a sturdy fellow not likely to panic under pressure. Joining them later would be several members of Fritz Wagner's entourage, who would line the tunnel to provide illumination (via flashlight) and calming voices as the escapees crawled past in the dirt.

Arriving at the site, the three were surprised to find a small crowd gathering, and this tableau: a few policemen, carrying machine guns; Fritz Wagner in his Mercedes; Harry Seidel in his smaller car; Dieter Thieme; Bernauer tunnel colleagues Wolf Schroedter (who considered this passage to the East a mere "rabbit hole") and Joachim Neumann (similarly worried about the outcome); an unidentified onlooker (possibly a journalist or a doctor or observer from the U.S. Mission); and an ambulance.

In preparation for this moment, the three tunnelers had packed a duffel bag holding an ax, a large saw, a hammer, several flashlights, two hand-cranked drills, and three weapons as well as a walkie-talkie to communicate with the West. They knew what the far end of the tunnel looked like deep in the East, but not what awaited them when they broke through into what (they guessed) would be the Sendlers' living room. The dwelling had no basement, no concrete to crack—just a wooden floor. Hence the drill and saw. They had been assured that the Sendlers knew they were coming and would be ready to flee, but they had no idea how long GDR guards stationed nearby would remain unaware of their presence. Hence the weapons: a Smith & Wesson pistol, an MG 42 machine gun from World War II, and a sawed-off shotgun. If nothing else they might scare off Stasi commandos without having to fire a shot.

Rudolph was eager to get started but also fearful. This was his first breakthrough. He didn't know the Sendlers and had no idea what their house looked like from the inside. He had little experience with weaponry. The arrival of so many refugees by truck seemed highly risky, and it might take two or three hours to get several dozen through a very tight tunnel, longer if one or more suffered (and no wonder) a panic attack. So many people now knew about the operation, with the refugee total seemingly rising by the hour, that the odds of Stasi penetration had multiplied. Even the nervy Herschel found the whole deal frightening.

None of them wanted to be a martyr, but the decision to take on this insanely dangerous project had been made, the date and hour set, and dozens of fellow Germans were on the way. They plunged ahead.

NEWS of the pending tunnel escape had reached the U.S. military in Berlin, prompting a cable to the State Department and other agencies from a lieutenant colonel, warning in its subject line of a *Possible Refugee Influx (U)*. The "U" presumably meant "underground." A "reliable source" had revealed that the influx would amount to about ninety escapees, in an operation organized by students and "butchers who operate trucks between East and West Berlin." CBS's Daniel "Shore," the cable advised, had earlier shot two rolls of film at the tunnel, to be aired August 13.

At 6:00 p.m., U.S. Mission chief Allen Lightner cabled Secretary of State Rusk. Someone at the Mission had spoken with an LfV operative—almost certainly Mertens—who confirmed that the escape operation was probably under way, as it had been set for 4:00 to 8:00 p.m. Mertens revealed that the LfV had known about the tunnel for some time, along with the identities of the organizers and CBS's plans to film. The LfV had "strongly advised" the diggers to "give up publicity [i.e., CBS] but group refused." Nevertheless, Mertens had examined the escape plan and found "intervention to stop project unwarranted."

Why not call a halt to the operation, given the State Department's insistence that the East Germans knew about the tunnel and were planning mass arrests? Mertens claimed there were "no concrete indications" that security was compromised. And besides: "Groups concerned have excellent record [of] successful escapes." This was a stretch, and the Bernauer team had no record at all. Still, Mertens had stated he was "unwilling [to] scrub whole project." If Mertens was the person the Mission had asked to warn the tunnelers of danger, it appears he did not do this; the only "warning" that had been delivered concerned the CBS coverage.

Lightner ended the cable on an upbeat note. He had learned

that Rusk had summoned CBS news chief Blair Clark to his office for another meeting at nine o'clock that morning, just hours after their late-night conference. Whatever was said then, squeezing CBS had apparently worked: "Schorr has ordered remaining cameraman scratch filming operation and leave Berlin for a few days," Lightner assured Rusk. But was this true?

WHEN Hartmut and Gerda Stachowitz exited the S-Bahn and reached the street near the Triftweg cemetery, they immediately sensed something was awry. A half dozen strangers, presumably fellow refugees, loitered in the area as minutes passed. After waiting for more than an hour, Hartmut and Gerda lost heart and started to hurry down the street, pushing a baby buggy, but saw their path blocked by menacing-looking men and big cars. Turning around, they spotted more men in trench coats emerging from behind trees and bushes. They were surrounded. Within seconds they were shoved into separate vehicles—Gerda with their child—and the cars sped off. All that remained of their dream: a baby carriage left behind on the empty street.

The Moeller and Schmidt families, who had also missed their trucks, were taking the final leg of their journeys to the Sendler cottage as pedestrians, with toddlers in tow. When they arrived separately at the Girrmann checkpoint along Puder Strasse, it was clear something was wrong. The signals to be displayed by *fluchthelfer* there were absent or wrong, and there seemed to be an unusual number of suspicious-looking characters nearby, wearing hats and trench coats on this August day. They looked like charter members of the Slouch Hat Mafia, as wary East Germans had dubbed Stasi agents.

The Schmidts quickly left the scene, but Anita Moeller argued with her husband that they should press on. Maybe they just got the signals wrong? Anita was overruled, and the Moellers, trying to blend in, joined the locals queuing at a fruit and vegetable stand in hopes of buying some just-delivered bananas, a luxury in the East. After a pause to calm their nerves, the

Moellers, too, quietly fled. Drinking wine until the wee hours and missing their truck might have saved them.

Also fleeing the scene: courier Dieter Gengelbach. He had come from the West to help load elderly refugees into one of the escape trucks near the Müggelsee. Then he had taken a taxi to Puder Strasse. After spotting a sign for the Sendler carpentry shop, he walked in that direction to see if anyone along the way had "Stasi" written all over them. Sure enough, he spotted three such characters and retreated. Gengelbach grabbed a taxi and headed for the nearest checkpoint to the West.

Mimmo Sesta, meanwhile, had accompanied his new truck-driving friend to meet a crowd of refugees at the appointed spot. After loading them in the back, the two approached a Girrmann checkpoint near the tunnel site. But they, too, noticed the crossed signals, so they turned around, without unloading passengers, and headed away from the danger zone.

The suspicious characters—those undercover Stasi operatives, now joined by three comrades—were still hanging out on residential Puder Strasse, waiting for more refugees to arrive. About fifteen suspects were already standing or strolling nervously nearby, awaiting instructions. The Stasi men had learned that the tunnel site at the Sendlers was now surrounded by their colleague Major Kretschmer, a Stasi commander, and his team. They expected an escape organizer to arrive at Puder Strasse any minute and summon the refugees, but when it got to be 6:30—and escapees appeared ready to melt back into the East—they decided to finally arrest them all. One of the Stasi agents expressed a single regret in his log: "Due to the confusing situation (and of the area) it is possible that some who wanted to escape to West Berlin were able to get away from our observation and could not be arrested." Perhaps he was referring to the very fortunate Schmidt and Moeller families.

Not so lucky was Manfred Meier. After his arrest he was marched a few hundred feet to the grounds of an old factory, an industrial yard enclosed by a wooden fence with a locked gate. At this police collection point, he came upon two to three dozen other arrestees, some with children, even babies. It was horrific. Many of the infants were crying. *This is a shit situation,* Meier told himself, thankful that he had not yet started a family of his own. After an interminable hour, a drab, unmarked bus pulled up, and they were all loaded into it to be taken to police head-quarters at Alexander Platz.

HASSO Herschel had no idea that his sister Anita, and many others, would not be arriving at the Sendler cottage. He and his cohorts, Uli Pfeifer and Joachim Rudolph, had dug up to the surface under the Sendler house, a messy job complicated by having to shovel through the hard coal slag installed under the house (this was used at many Berlin homes lacking basements as weatherproofing for the winter). Now it covered the diggers in coal dust. Never mind. They had to make a large hole in the thick floorboards from below, wider than an overweight human, using basic tools, in dim light, with coal dust in their eyes and little air to breathe.

First they marked off a rectangle on the underside of the planks and drilled holes in each corner with their hand-cranked tool. Taking a small keyhole saw, Herschel and Rudolph took turns cutting away at each side, with Pfeifer below them han-dling a flashlight. It seemed to take forever, requiring so much muscle power they had to trade off every minute or so. Now it was sawdust falling in their eyes. They were also making con-siderable noise and figured the Sendlers, waiting above, must have heard them by now.

After about ten minutes, and only halfway through this task, they suddenly heard a woman scream, "Go away! Get out of here!" This had to be Edith Sendler, but why was she saying this?

Hasso, under the floorboards, shouted a reply, "You can come with us, don't worry, it's okay!"

"We don't want to come to the West!" the woman shrieked. "Not at all! Get out! Get lost!"

Hasso started offering her money, several thousand DM, to shut up and get ready to exit the East as she and her husband had allegedly promised. When she again shouted *nein*, he asked her to at least leave the house and remain silent for a couple of hours until the escape operation was over. He received no reply. Perhaps she was complying? The house grew quiet. The three men in the tunnel, in a state of agitation, with adrenaline pumping, quickly conferred. They decided they had no choice but to finish the job. Three trucks of refugees should soon be arriving.

AT that moment, Friedrich Sendler was talking with the two Stasi men who had returned to his backyard fence and were now no longer asking but demanding entry. Sendler told them he didn't like strangers "loitering" near his yard, and that he was already nervous because some people across the Wall seemed to be studying his property. The Stasi agents told him to open the damn gates. Then they moved to the front of the house for a look around. If Sendler had ever given permission for use of his house as an escape hatch, he now knew that idea was dead.

The Stasi operatives were still conversing with Herr Sendler when his wife suddenly emerged from the house. Perhaps she was contemplating going for a long walk or drive with her husband while the tunnelers made use of their house, as Hasso had requested. But now, here, a surprise: the Stasi! She complained to them about feeling ill, likely a stalling tactic. The two agents failed to depart, and indeed perked up, for Frau Sendler seemed nervous. Asked again why she had left the house, she now responded, as the MfS report put it, "that she was in her living room and that there were drill holes in her floor." She said nothing, however, about her excited exchange with Hasso, nor his request that she leave the house for a few hours.

Without further delay, two Stasi agents plus Edith Sendler entered the house on the right side, stepped softly through the vestibule, along a narrow corridor, and "quietly went into the mentioned room" situated in the middle of the ground floor, according to the Stasi log. There, near the front window, they confirmed that floorboards had indeed been "penetrated" by a drill in several places, and noticed wood shavings on the floor. The three retreated to the vestibule just inside the front door. Then, more quietly still, they returned to the living room. This time they heard sounds below that indicated several tunnelers were about to break through the boards. The agents heard one tunneler say, "We have found the right house." Another seemed to hiss into a radio, "My pistol may not fire—bring the machine gun!" Moments later, noises from below suggested that weapons were being gathered and loaded.

The Stasi men returned to the vestibule to await the arrival of soldiers bearing Kalashnikov rifles. Major Kretschmer, now in the house, ordered them not to burst in and make the arrests until all the perpetrators had occupied the living room. He had personally witnessed Stasi commandos opening fire on Heinz Jercha back in March so he would not likely flinch if shooting broke out here.

ACROSS the border fence, high in the abandoned railyard building, Piers Anderton and his NBC cameraman watched the disaster unfolding. From the tower's window he had a clear view across the tangle of bushes leading to a wooden fence and several lines of barbed wire (which comprised the Wall here) to Kiefholz Strasse, a sidewalk parallel to the street, and then the Sendlers' front yard and cottage. Elevated train tracks crossed the sky just to the left. One block to the right of the house, Puder Strasse led to Treptower Park several blocks in the distance. Occasionally GDR policemen or guards had marched down the sidewalk that afternoon, their caps pulled down and long coats blowing in the unusually chilly August breeze, past the Sendler house. Anderton wondered whether the frequency of these patrols was typical. He had no idea if Daniel Schorr and his CBS cameraman had arrived below his perch.

Opening a curtain for brief moments, the NBC team filmed

the guards striding past the Sendler cottage, often taking a nonchalant glance or two at its front door. Did they know something was about to happen there? Anderton directed the camera to a man in a white shirt crawling through the thick bushes on the Western side of the fence. Tunnel organizer? Police? Journalist? Or CBS cameraman?

Now, in the early evening, he saw a dark-haired, heavyset woman exit the target house. Two men in trench coats emerged, seeming in no great hurry, from the Sendlers' backyard. The woman, maybe fifty years of age, was now chatting with a third visitor outside her house. A man about her age, possibly her husband, stood nearby. The pair in trench coats approached the steps leading to the Sendlers' door at the far right and entered the house there. Curtains were still drawn on the larger room to the left.

Now two men in uniforms, soldiers or guards, led by a civilian, no doubt Stasi, arrived, armed with Kalashnikov rifles. Someone inside must have ordered them to step lightly so they could sneak up on the tunnelers, for the soldiers carefully put aside their heavy weapons for a moment and took off their boots, like children admonished by their mother on a rainy day, before slipping inside. Whatever happened next was hidden from the NBC camera. Anderton might have wanted to shout out a warning to the three tunnelers, but it never would have reached them.

THE three diggers, unaware of the trap being set a few feet above, had to hurry now. They sawed a couple more inches along the floorboards, and then, to finish faster, knocked the wooden rectangle out with an ax. This caused quite a racket—risky since they were just yards from a street patrolled by guards. Even their comrades across the border could hear it. Someone shouted down the tunnel, "Quiet! Stop it!"

Finally they pushed the wood through. Hasso Herschel took out a small mirror and raised it above the floor. No one visible. He climbed out. Joachim Rudolph followed. Uli Pfeifer pushed the kit bag through the hole and joined his comrades above ground. They saw that their hole was less than five feet from the side of the living room facing Kiefholz Strasse. What a fright they would have given any refugee or soldier: their faces were still covered with black soot. Joachim took out his pistol and walked to one side of the large, dark room. There was a table in the middle, a few chairs, a couch, a TV set. To the rear: more rooms across a corridor. But still no sign of the Sendlers.

Curtains covered the windows, certainly a good thing. Rudolph pulled back a corner of one near the escape hole—and spotted a man skulking around the yard. He was not wearing a uniform or waving a gun, so Rudolph was unsure if he was Stasi, refugee, or *fluchthelfer*.

Inside the house, Stasi operatives, joined by two shoeless soldiers with automatic rifles, remained huddled in the vestibule. Given the large number of armed agents and military personnel in the area, it was a surprisingly small force. Stepping lightly into the corridor they could peek into the living room through a partly opened door. Around 6:45 they spotted a man popping out of the hole, but they could only see the front part of the room. Perhaps the agents failed to notice the other two tunnelers, or they expected several more to emerge, heavily armed. In any case, the Stasi and the soldiers waited. "At 7 p.m. the work [on the hole] had been finished and voices could be heard," the Stasi log recorded.

At that very moment, warnings from the West boomed over the tunnelers' walkie-talkie: "Come back, come back! There are people on the property!" They clearly did not mean escapees.

Across the border in the West, a worried Joachim Neumann heard shouts from the hill behind him: "Stasi!" Fritz Wagner, sitting in his Mercedes, added his own frantic plea, honking his horn in alternating long and short bursts, but no one inside the house could hear it. West Berlin police at the tunnel entrance readied their weapons. They had been ordered never to fire across the border, but a few warning shots in that direction might buy the tunnelers some time.

Yet the three tunnelers did not take flight immediately. They carried serious weapons, but they had never fired them and did not even know if they would work after many minutes in a soggy tunnel. Still standing in the living room, they heard the warning shouts from the West down the tunnel and over the radio growing louder. After a brief conference Hasso made the final call: "Okay, we stop!" Swiftly they packed up the duffel bag and dropped into the passage below and crawled as fast as their arms and legs would take them back to the West.

At 7:10, the Stasi log observed, "complete silence had been noticed from the tunnel." Stasi operatives and barefoot soldiers finally entered the living room—and found their prey gone. Fearing massive "terrorist" firepower, the Stasi had waited too long for military backup, and then dithered some more, countering their reputation as highly organized, competent, and quick to act. All they could do now was arrest the Sendlers and note that two rooms in their home were suspiciously "overloaded" with black market goods: butter, sausage, flour, wine, cognac, champagne, bathroom tissue, coffee, and chocolate. They recovered an ax from the living room floor and, down in the hole, two drills.

WHEN they climbed out of their cavern back in the West, the three tunnelers did not pause to ask anyone there what had happened, though they were already haunted by the question: *When was the last time anyone cleared this mission with the Sendlers?* Had Harry Seidel unwittingly led them into a trap? Dirty, bloody, exhausted, and feeling lucky to be alive, they jumped into a police van, which sped off. The cops spotted the cache of weapons in the duffel—all of them illegal in West Berlin—but did nothing.

At the tunnel entrance, most of the onlookers cleared out in case the frustrated East Germans decided to take target practice across the border. It would be nice if no one beyond this small circle knew that any of this had happened. For several hours, however, cops and civilians in the West came and went at the site. Some of them studied the Kiefholz/Puder Strasse area through binoculars. The Stasi watched them right back, even taking down the license number of a West Berlin police car. Fritz Wagner returned in his black Mercedes for another look. The final entry in the Stasi report that night read: "At 1 a.m. the tunnel has been secured by own forces."

So ended an historic confrontation. For the first time that year a single operation had brought together in one place, just a few hundred feet apart, nearly the entire cast of characters absorbed in the Berlin escape drama: tunnelers, couriers, and refugees; Stasi, GDR guards, and soldiers; West Berlin police and American military; and journalists from one, if not two, American TV networks.

Despite a badly blown operation, the three tunnelers, as well as the three Schmidts, the three Moellers, and the fiancées of Meier and Sternheimer, had lucked out, narrowly avoiding arrest or worse. Dozens of others were not so fortunate. More than forty refugees had been arrested. Some were quickly induced by Stasi interrogators to name people who had assisted them. Sternheimer, betrayed by Uhse earlier that day, was arrested at the Heinrich-Heine checkpoint that night when he tried to reenter the West. Taken to police headquarters, he thought he might talk his way out of it—claiming he just happened to be in the East as a student and tourist—until he learned that his fiancée Renate had been arrested at home.

The Stasi wasted no time with Friedrich Sendler, starting a marathon interrogation session at 10:20 p.m. that night with, "Which offenses have you committed against the GDR?" Sendler said he knew of none, adding that he was "outraged that I have been entangled in a matter which I have nothing to do with. I am very sorry that an underground tunnel was dug toward my property, which I had no knowledge of until today, otherwise I would have informed the border police." Sendler claimed that he had stayed in the carpentry workshop all day, supervising six employees, without stepping foot in his house. "Around 6:15 p.m., my wife who was in the kitchen heard sounds from underneath the house," he said, "and immediately came to the entrance door of the veranda where the two persons and I were standing and she said that she heard sounds coming from under the earth." This omitted a key detail: that she had *not* relayed this information right away, first complaining of feeling ill.

The interrogator wasn't buying it. "Your statement is illogical," he informed Sendler. "How do you explain the fact that the intended tunnel, originating in West Berlin, ends in *your* home of all places?"

Sendler replied, "Neither now nor in the past did I have any contact with persons who would ask such a thing of me." Then he offered a brave, if foolish addendum: "Even though I do not completely agree with the actions and measures of the GDR, and I inform myself with West German television for opposing views from time to time, I would never be willing to commit such an offense." Clearly he wouldn't be going home any time soon.

Edith Sendler was also questioned through the night. She stated that, since the Wall went up, she and her husband had not been visited by a single person from West Berlin nor any "middlemen" in the East. She had been "extremely shocked" to see the invader's drill poking up in her living room. Then, she said, she had dashed outside and "immediately informed" the two Stasi agents (a sad lie). And finally: she had no explanation for all those scarce goods, the "luxury items" and alcohol discovered in their home. "I realize that my words are believed not to be true," she said as the questioning concluded, "but I can't make any other statements about it."

Faring even worse than the Sendlers were Hartmut and Gerda Stachowitz. They had been taken in separate cars to separate prisons—the women's prison for her, brutal Hohenschönhausen for him. As Gerda was interrogated, her infant son was placed on a table nearby, his cotton diaper soaking wet. Around midnight he was taken from the room, with no explanation of where he was going, and if and when his mother would ever see him again.

9

Prisoners and
Protesters

AUGUST 8–14, 1962

The quality of U.S. intelligence on the botched Kiefholz operation left something to be desired. Following the message sent by Berlin Mission chief Lightner around noon on escape day, Secretary of State Rusk did not hear anything about the escape for many hours. At 11:15 p.m. that night, Lightner finally cabled a brief update: "Escape effort not made today. Not clear at this point whether project abandoned or postponed."

Rusk received no further word until the following afternoon, when Lightner sent him another cable. The exodus had apparently been "called off" after a "counter-surveillant in East Berlin observed heavy concentration of uniformed personnel" near the Sendler home. The LfV, presumably via Mertens, had informed the Mission that as many as twenty refugees had been arrested. Several dozen others had evaded arrest after a "warning message" was allegedly "sent to participants." An unnamed organizer had told the LfV that the tunnel would not be used again—surely the understatement of the month. The cable concluded:

Worst aspects of what could have been politically very embarrassing development seem to have been avoided. However, with two known arrests and possibly more of

would-be-escapee group, it is possible SED regime may
mount show trial with statements from those arrested im-
plicating CBS participation in planned operation. There is
also possibility West Berlin organizers may eventually use
Schorr, O'Donnell involvement to excuse failure of their
operation, especially if as many as 20 actually arrested and
brought to trial.

A State Department staffer forwarded this cable to Mac Bundy
at the White House, with the note that Rusk "thought the President would want to see this. Salinger is out of town so he has not
seen."

Also on August 8 an unidentified member of the American
diplomatic staff in Germany (most likely Lightner) wrote a
lengthy "Memorandum for the Record" with the title "Compromised Escape Tunnel." There was one bizarre new detail: Harry
Seidel's friend Rainer Hildebrandt "had obtained and turned
over to one of the Western participants a VoPo uniform which
was to be used in East Berlin to provide security coverage for
the escapees when trucks transporting these persons reached the
eastern parking lot." The memo carried no claim by Mertens
that he had warned the tunnelers that their project had been
"compromised," as he had done for projects earlier that year.

The only person known to have been alerted to high danger
surrounding the tunnel was Daniel Schorr, and he still didn't appreciate that warning one bit. An embassy staffer in Bonn asked
Allen Lightner to have another chat with the CBS correspondent
to make sure he was aware of the "results and implications" of
the failed escape attempt. They had heard that Schorr's current
"line" was that the State Department can't seem to "take action"
to aid those seeking freedom in Berlin "but it can certainly prevent it." Dean Rusk approved the Schorr meeting in a cable to the
Mission (again with a copy to Salinger and Bundy). He added:

US officials have no apology for prompt action taken with
CBS and Schorr; Schorr involved himself in a matter which

was far beyond his private or journalistic responsibilities and proceeded amateurishly in a matter filled with the great-est danger for all concerned. As we anticipated, other side turned out to be fully aware of entire matter and laid a trap which could have resulted in massacre [of] those involved. We cannot help but be dismayed at degree of involvement which Schorr had contemplated, apparently without any re-gard for grave possible consequences.

Of course, the "other side" had not become "fully aware" of the tunnel until the morning of the operation (via Uhse), so it was hard to say why State had anticipated this with such certainty.

THE Stasi had blown a golden opportunity to arrest three of the leading Berlin tunnelers inside the Sendler cottage, but they did have a chance the following day to study the criminals' work. They didn't dare crawl too far into the narrow cavern, how-ever—it was that dangerous looking. Even the search dog they brought to the site wouldn't go farther than about thirty feet in. After inspecting the tunnel, many of the soldiers and Stasi agents helped themselves to the Sendlers' upscale possessions. The couple seemed headed for long prison terms; they would hardly need the cosmetics, fine nylons, woolen sweaters and jackets, high heels, expensive jewelry, and watches. The Send-lers had probably obtained them illegally from the West any-way, and everything likely would be confiscated by the state. Why shouldn't the frontline "protectors of the border" share in the bounty?

A few hours later, the Stasi completed a detailed report on the tunnel escapade. They lauded their own "protection of the peace" against the "west Berlin terrorist organization Girr-mann" (while failing to admit they had bungled the arrest of the perpetrators). They credited information that came directly from Siegfried Uhse, and detailed the names and addresses of those arrested—forty-three so far. The arrests were listed by site: six each at the two truck pickup locations, nineteen near the tunnel on Puder Strasse, plus the courier Wolf-Dieter Sternheimer and eleven others detained "independently" off-site by border police. More arrests were expected as interrogations continued.

The report detailed Sternheimer's work with Girrmann on plans for "violent border breakthroughs" going back many months. He was one of the few who had known where the escape house was located, as shown on the map he had entrusted to Uhse. Also featured was "The Accused Stachowitz, Hartmut." Since July, Stachowitz had been in contact with "the American student" at the Girrmann office—a reference to Joan Glenn—concerning the smuggling of his wife. The Stasi had questioned more than thirty other detainees so far. One of the accused could not be interrogated "because of a heart attack and was sent to the VP hospital."

Manfred Meier, meanwhile, felt he had a chance for release, even though Stasi interrogators at Hohenschönhausen kept telling him, "We know everything." Meier continued to proclaim his innocence; he had merely been in the neighborhood of the escape to visit a cousin. And he really did have a cousin in the area.

Thanks to his warning on Puder Strasse, his girlfriend, Britta, had avoided arrest. Now, with great trepidation, Britta decided to tell her parents about the thwarted escape and the arrest of her boyfriend. She sat on the edge of her father's bed and said, "I have to tell you something." Her father took her hand, and replied, "You know, Britta, I would have done exactly the same thing. Don't blame yourself for anything." A few hours later, after someone in detention revealed her name to the Stasi, she, too, was arrested.

Besides extracting the names of plotters still at large, the Stasi hoped to use the mass Kiefholz arrest to find out about other tunnels in progress. (Those involved in the Bernauer tunnel recognized this threat immediately.) Sure enough, the interrogation of one refugee, a painter, produced a vague but troubling tip. His contact for the Kiefholz escape operation was named Gengelbach, he said. This Gengelbach had told him that there were two separate West Berlin tunnel groups. One of them was made up entirely of students. Gengelbach did not know who was sponsoring the students, but he was sure they were "working on another tunnel which already is 150 meters long and is supposed to be finished at the end of August or beginning of September." The details were sketchy but the Stasi now knew to be on the lookout for any fresh information about such a major project.

UNLIKE the MfS, the Girrmann Group remained perplexed about what had gone wrong at Kiefholz. The morning after, they huddled with a large group including Harry Seidel and Dieter Gengelbach. Clearly an informer had tipped off the Stasi, but after a heated discussion, no one could identify a likely suspect.

The following evening, Siegfried Uhse was summoned to the House of the Future to review the tragic events. One imagines that "Hardy" felt a bit nervous en route. By then his Stasi identity could have been exposed in any number of ways. When he arrived, he encountered all three leaders: Girrmann, Thieme, and Köhler. Not a good sign. Köhler welcomed him with "Another corpse is here." Uhse's stomach no doubt dropped—until he learned that this only meant he was stuck in West Berlin forever, because it would not be safe for him to go to the East, as now the Stasi might be on to him.

The three organizers asked Uhse and other couriers—the ones not arrested—to review what they did and saw on August 7, as they searched for clues about who the informer might be. Uhse recalled that day in detail (leaving out the part about handing their map to the Stasi), including his meeting with Sternheimer and Stachowitz at one in the afternoon. But since those two had both been arrested it seemed unlikely they were the snitches. A Girrmann staffer told Uhse he should count his blessings—he was "lucky" he had not been busted in the East himself.

Another visitor arrived at the office, a heavyset man in his forties, with dark, graying hair and glasses. Uhse learned that this was one of the group's chief financial backers. Thieme told the man that the August 7 disaster was the greatest setback since the Wall had risen. The visitor had a more positive message. He had just come from Bonn, where he had met with top businessmen who offered to provide up to 2000 DM for each refugee the Girrmann Group freed—the highest amount if the escapee was a student or skilled worker. What persuaded these funders? "I explained to them how they can deduct this from their taxes," the emissary, who was probably a government official, explained. "If they can deduct something from their taxes—then I hold them in the palm of my hand!" To receive the money, all Girrmann staffers had to do was to ask each refugee to make a recording about his or her background; there was no need to reveal how they were smuggled to the West. The visitor from Bonn urged fast action so the money could start flowing.

On the down side, he continued, officials in Bonn would no longer offer assistance to the Girrmann Group, because they felt such aid was "compromising" the government. They demanded that Köhler resign his position managing the House of the Future, as they held him responsible for several failures, especially the Kiefholz fiasco. The visitor also revealed that another major tunnel—"in the north" (this would include the Bernauer area)—was in progress. This now seemed an open secret. Who else knew?

A discussion followed concerning the missing courier Manfred Meier. The ever-helpful Uhse offered to leave immediately and talk to Meier's brother. When he returned, he said the brother seemed to be hiding something. This was Uhse trying, with some success, to focus attention on the missing courier as the likely snitch. (No one in the West knew that Meier had been arrested.) Uhse then crossed the border and reported all of this to his MfS handlers.

They noted that Uhse felt confident that he had now "gained or consolidated the trust" of the Girrmann leaders. So the Stasi piled on assignments: try to learn more about smuggling via cars; obtain the names of remaining couriers; find out more about that visitor promising money from tax-dodging businessmen; help Köhler in the "dismantling" of the House of the Future office "and try and get access to the material there," especially the list of East Germans applying for escape help. And: "try to discover the location of the new tunnel and find out who builds the tunnel."

At their next meeting, Uhse reported, "I have been informed that tunnel is about 150 meters long. Nothing is known about the entrance, also not if it is being worked on." His minders pressed: *Where is new tunnel? As soon as knowledge is gained send info via telegram.*

ON the afternoon of August 10, Daniel Schorr was called on the carpet at the U.S. embassy in Bonn. He now learned, if he had not known already, about the many arrests in the Kiefholz tunnel operation, and that a trial might implicate him and his network. Schorr heard his personal involvement labeled "incredible." He was pushed to admit that he had paid more than 5000 DM to the tunnel organizers. Why? *Something* had to be done to dramatize the horror of the Wall, he replied, since the State Department had taken a "negative attitude" toward commemoration of the anniversary. A mass escape, with network coverage, would have served that purpose. Schorr said that his chief aim now was to find out how the plot had been compromised, but the embassy official ordered him not to discuss it with anyone except officials at the Mission.

A summary of the conversation was sent by an embassy official to Rusk and Lightner, and then on to Salinger and Bundy at the White House. It claimed that Schorr "appeared chastened by the fact that a plan which was to be his greatest television achievement has failed." Yet he "did not give the slightest appearance of being contrite." One brief passage in the summary raised more questions than it answered. The embassy official had told Schorr that the tunnel plot could have resulted in a large-scale loss of human life "had effective action not been taken to prevent its fulfillment." What "effective action" (besides Uhse's) had been taken and by whom?

The following day, Schorr visited Lightner at the Mission in Berlin over on Clay Allee. The conversation followed the lines of the Bonn meeting, but Schorr added this complaint: he had heard that Piers Anderton of NBC was spotted at the scene of the tunnel entrance in the West and had filmed police activity and the bust at the Sendler cottage. NBC would likely use some of this film (which actually amounted to only two minutes of footage) on an upcoming news program, Schorr presumed.

"Schorr concerned that competitor had done what CBS through [State] Dept. intervention had been prevented from doing," Lightner cabled Rusk—once again with a copy to Salinger at the White House. He added that "while we hardly in position assist CBS competitive problem, we are concerned how best to prevent similar actions in future by either CBS or NBC." This was vital as they were "not at all sure Schorr will not attempt film another escape" though he "may avoid dangerous advance arrangements directly with organizers." The State Department may wish to consider "high-level intervention with NBC along general lines taken with CBS. In view Anderton's generally uncooperative attitude we will initiate no action here with him unless so instructed."

Well, they had certainly read Anderton right. But Schorr and CBS had just begun to fight—no longer *for* their film, but *against* NBC's.

PRESIDENT Kennedy's long-simmering unhappiness with U.S.–NATO nuclear strategy in Europe reached a climax in August when he summoned top advisers for a nearly ninety-minute meeting in the Oval Office. Also in attendance was Walter Dowling, ambassador to West Germany. It was seventeen years to the day since the last time the United States had used a nuclear weapon, killing at least 75,000 Japanese in Nagasaki.

Nuclear war continued to hover over every discussion of Berlin. When he came to office, Kennedy had discovered that his predecessor had not fully mapped out possible U.S. and NATO responses to a conventional Soviet attack in Germany short of what was labeled "massive nuclear retaliation." He was also worried about the possibility of an accidental nuclear war sparked by misread signals, and concerned that the line of command gave top generals the authority to launch missiles. When he asked top military aides, "I assume I can stop the strategic attack at anytime. . . . Is that correct?" the answer was often unnervingly vague.

Some of the generals, he felt, spoke rather cavalierly about the effects of nuclear war. A top Strategic Air Command general, briefed on the long-term genetic effects of radioactive fallout, had quipped, "It's not yet been proved to me that two heads aren't better than one." Unlike many of his advisers, Kennedy felt there was too much emphasis on who would win a nuclear war, too little on the survival of the human species.

Nevertheless, the President had affirmed that it was U.S. policy to initiate the use of nuclear weapons if the Soviet Army invaded Western Europe. A U.S. target list, titled "Atomic Weapons Requirements Study for 1959" and prepared by the Strategic Air Command, included not only thousands of sites within the Soviet Union but also ninety-one in and around East Berlin. Besides several Soviet air bases in the suburbs, dozens of sites within the city made the list as part of its "systematic destruction": factories, railroad hubs, power stations, radio and TV transformers. There appeared to be little recognition that hitting even one or two sites in East Berlin would produce firestorms and radioactive fallout certain to reach West Berlin. One haunting entry, simply and unapologetically labeled "population," took dead aim at the civilian center.

President Kennedy, still a young man and with two small children, often spoke privately of the dread he felt when contemplating nuclear attack. When his brother had met him at the White House after the unnerving summit with Khrushchev, Robert Kennedy noticed tears on his face, the first time he had ever seen him cry over political stress. "Bobby, if a nuclear exchange comes," JFK confided, "it doesn't matter about us. . . . The thought, though, of women and children perishing in a nuclear exchange, I can't adjust to that."

Henry Kissinger, an ambitious young Harvard professor whose family had fled Bavaria after Hitler rose to power, was among the consultants pushing for a layered nuclear-response plan, particularly in response to a Berlin crisis. Kissinger was a hard-line anti-Communist but, unlike many others, favored a little nuance in confronting the Soviets. He had encouraged Kennedy to order a full review of U.S. nuclear first-strike options. Unless the United States knocked out every Soviet missile site—which was unlikely—the enemy would fire off all of its rockets in response, and tens of millions of Americans would perish. This left the President a choice between what policy insiders liked to call "surrender or suicide."

Now Kennedy wanted to know whether there was a way to go nuclear yet limit the targets—and the casualties on both sides. He favored a "flexible response" to a Soviet attack across the German border rather than automatic "massive retaliation." The pages of paper generated at the Pentagon were so voluminous that the proposed new strategy became known internally as "Horse Blanket." When trimmed, the name that stuck was "Pony Blanket." Cut some more, to just a few pages, and you had the misleadingly cute "Poodle Blanket," approved in October 1961 as document NSAM 109. It projected a "sequence of graduated responses to Soviet/GDR actions" leading, over time, to limited and then (if necessary) massive nuclear attack. It could be a reprisal to a Soviet nuclear action—or it could be a first strike. A top Kennedy aide referred to that as "the moment of thermonuclear truth."

A few days before the August 9 nuclear-policy update at the White House, the Soviets had resumed nuclear testing with a bang—a 40-megaton shot over an Arctic island, judged to be the second biggest explosion ever achieved. Kennedy's hopes for a test ban faded even further. On top of that, West German defense minister Franz Josef Strauss had just publicly criticized the United States for allegedly aiming only at military targets with its nuclear missiles, neglecting civilian ones. (The administration quickly assured him that America aimed at both.) Strauss also called for NATO to arm its frontline forces with "battlefield" nuclear weapons, claiming use of them would not necessarily provoke a deadly Soviet response. The Americans opposed this idea thoroughly.

Still, as the high-level meeting at the White House proceeded, Kennedy admitted candidly that the Soviets had the Americans over a barrel in Berlin. If the Soviets seized Berlin, isolated as it was, should the West go nuclear? Or wait until the enemy crossed the border into West Germany? Or not even then? In other words, were the allies just bluffing about responding to any of these moves by using nukes? If they did use them, would their assault be limited or extensive?

John Ausland, deputy director of the Berlin Task Force, briefed JFK once more on "Poodle Blanket." Dean Rusk commented that what worried America's NATO allies was that if the West didn't go nuclear in a showdown, "the Soviets will grab off a considerable part of Germany and then want to negotiate from that point." Robert McNamara, however, believed U.S. allies were operating under the misconception that "there is a *salvation* in tactical nuclear weapons," partly based on a "lack of understanding of the use of tactical nuclear weapons and the result thereof." The deadly effects of radiation drifting to earth all over Europe was somewhat underappreciated. McNamara could have mentioned that modern warheads were far more destructive than the Nagasaki bomb, but did not.

Kennedy insisted that the first-strike threat would have more validity "if you didn't have the problem of Berlin." For one thing, American forces there would be trapped fighting in East Germany behind the nuclear line. Also, when it came down to it, would the allies really start a nuclear war over already divided Berlin? Bundy brilliantly put the entire debate over a first strike with nuclear weapons in simplest terms: "It's only a good thing to do—if you're never going to do it." Little more could be said after that and, indeed, the meeting quickly wound down, nothing resolved.

IT wasn't hard for a young man like Peter Fechter to imagine a better life for himself beyond the Wall. He was the only son in a family that struggled to eke out a living in the shattered postwar economy of East Berlin. His father worked as an engine builder and his mother as a sales clerk. The spirited blond teen, who had left school at fourteen to apprentice as a bricklayer, had relished the visits, before the Wall, to his sister Liselotte in West Berlin. She had left the East in 1956 to marry. Peter, who turned eighteen in 1962, dreamed of joining her. Having enough money to move out of the three-room apartment he shared with his parents and sister in the Weissensee district seemed impossibly far off.

"Everything is going wrong over here," he wrote in one letter to Liselotte, complaining that "those swine" had again stepped up quotas that required working longer hours for the same paltry pay. It was bad enough that months earlier he feared he would be forced to helped build the Wall. "This morning I went to Friedrich Strasse [near Checkpoint Charlie] with a friend and saw the American flags. . . . So near and yet so far! I felt like crying. I divide my time between the building site and my girl friend. . . . We get along well together, but her mother thinks we should wait two years before we get married. Who knows what will have become of us by then?" He was receiving excellent reviews at his job site, but surely in the West he would have a better chance of hard work paying off.

With his equally frustrated friend Helmut Kulbeik, a mason who worked with him rebuilding the former Kaiser Wilhelm Palace on the Unter den Linden, Peter began to think of escape. Their building site was not far from the Wall. That summer they used their lunch breaks to scout locations along the border grounds, near and far, though they had yet to devise a detailed plan. Neither mentioned any of this to their families. One early August day they discovered a run-down factory near Checkpoint Charlie that housed a carpentry workshop directly on the border strip. Its back windows faced Zimmer Strasse, providing a short dash to the Wall—if one were willing to dare it.

AS the first anniversary of the Wall neared, American and German leaders grew nervous about how it would be commemorated on the streets. West Berlin officials had announced plans to keep recognition calm and limited, marking the day with a few minutes of silence at noon when all work and travel would stop. A "freedom bell" would toll and Mayor Brandt would offer a few words on TV and radio. West Berliners were asked to stay in their homes that evening and write letters to friends and family in the East. Charles Hulick had cabled Dean Rusk that the observance was to be low-key "and with minimum possible opportunity for demonstrations which might lead to incidents. . . . Special precautions being taken on both sides of wall should prevent development large-scale actions."

Some legislators and media commentators in West Germany called this far too passive. Various groups planned to lay wreaths and raise voices at the sites where Easterners had been killed trying to flee. One popular tabloid called for church bells to ring, sirens to sound, and car horns to blare, for "we must scream—we must alert the world." The Communist bloc had stepped up its propaganda campaign, charging that West Berlin was an espionage center and source of tension. The CIA warned in its daily briefing for the President that the East German military seemed to be rehearsing—"for precisely what we haven't been able to tell." And Khrushchev might be planning "a major gesture."

When August 13 arrived, the day was marked in the United States with a ninety-second program on the four leading radio networks: General Lucius Clay spoke and the Liberty Bell in Philadelphia rang out for freedom. (Still, General Clay bitterly complained to the State Department that it was severely downplaying the anniversary.) For whatever reason, NBC did not choose to air its dramatic clip of the recent Kiefholz bust. In West Berlin, both the official and unofficial events went off as hoped. The midday stoppage and silence were widely observed and, many felt, profoundly moving. Then the honking of horns proved deafening. When a large cross labeled *We Accuse* was erected along the Wall near Wilhelm Strasse, the East Berlin police tried to knock it down with water cannons. Police on both sides of the Wall were soon exchanging tear gas volleys.

Another action signaled that protests might take a more militant direction. The Girrmann Group had organized about a dozen teams to board elevated S-Bahn trains and pull the emergency cords, halting them between stations during the moment of silence. Even though the trains ran in the West, the S-Bahn was still owned and operated by the East (in theory, even the tracks were Communist), so this would send a message across the border. Dieter Thieme asked Wolf Schroedter to take part in the protest, and Schroedter invited Joachim Rudolph to join him. Their target station was in Moabit. It was an act of civil disobedience—they pulled the emergency cord and then stood next to it, awaiting the East Berlin railway police (who, annoyingly, rode the cars even in the West) to arrive. Asked for their IDs, the two men refused. They were turned over to the West Berlin police and given a choice of spending five days in jail or paying a 28 DM fine. It was easy to choose the fine since the Girrmann office covered it.

By late afternoon, the protests, as officials feared, had turned increasingly angry. Brief notations on the duty officer's log of the U.S. military's Berlin Brigade captured the mood:

> *6:55 p.m.*: Received report from C/P Charlie that the Sov. War Memorial bus at corner of Koch and Freid ... driver was hurt by stone thrown by someone in crowd.

> *8:30*: 2 mile long line of cars protesting toward bridge at 942182.

9:55: Garten Platz—BSP [East Berlin police] throwing tear gas grenades—West Berlin police throwing grenades back—500 people in area.

10:47: 1000 sit-down strikers moving from Bernauer towards Kreuzberg.

10:53: 2000 demonstrators at Moritz Platz and Prinzen Str. Strong West Berlin police force pushing them back away from the "wall."

10:55: About 20 cars broke thru barrier at the Brandenburg Gate and went as far as the British observation platform.

Midnight: Oberbaumbrücken, approx 100 people attempted to cross bridge into East—West Berlin police put trucks across entrance to bridge—resorted to using night sticks.

West Berlin police reports recorded many other acts of violence and what they termed "near riots." Protesters had tossed stones and beer bottles over the Wall at guards. In one spot they shattered streetlights with stones on the Western side so they could heave rocks into the East under the cover of darkness. A Bernauer Strasse crowd of three thousand tried to break through the Wall—then stoned West Berlin police who turned them back. Four policemen were injured there and twenty more elsewhere.

That evening, there was a chaotic scene across from the vacant, and now infamous (in certain quarters), Sendler cottage. A huge force of East Berlin military and police had been deployed throughout the area after four men were spotted less than a hundred feet over the border fence in the West pointing a movie camera at the Sendler home, likely either a TV or newsreel team. Someone from the West had fired a shot or shots at guards in the East. Others threw stones. Armed West Berlin police appeared and scanned the East through binoculars.

Then the rowdies in the West started shooting out streetlights between Kiefholz Strasse and Puder Strasse, site of most of the August 7 arrests. Shots and stones aimed at GDR guards continued until just past midnight.

"Due to the actions and behavior of the enemy, it is expected that a border breakthrough had been prepared," the East German captain's five-page report stated, also citing the presence of cameras. The area across the border from the Sendlers' would now be watched "permanently." West German police near the site must be monitored closely, for they clearly aimed to help "border violators." Inspection of basements, garages, and workshops on the Eastern side, presumably in search of more tunnels, would be carried out by the most trusted military squads.

The next day someone wrote by hand at the top of the formal report: Harry Seidel's Kiefholz tunnel leading to the Sendler property had "been exploded by border police."

10
The Intruder

Despite their close brush with catastrophe—nearer to arrest than success—Hasso Herschel, Uli Pfeifer, and Joachim Rudolph returned to Bernauer Strasse eager to get to work again. Putting their terrifying experience behind them, the three young men felt confident that their own operation was far safer than what they had just barely survived. This ignored the fact that their tunnel had not yet penetrated very far into the hostile East, meaning they had many weeks of burrowing yet to do. They could be hit with another massive leak at any time and, as for security, so many diggers and Girrmann Group associates now knew about the project that one had to wonder.

Even as Herschel, Rudolph, and Pfeifer picked up their shovels again, they remained mystified about what had happened in the Sendler cottage. Rumors placed the Sendlers—or Wagner and Seidel—at the center of the disaster, as the tunnelers had no way of knowing that a Stasi snitch had doomed the plan anyway. They had been assured by Harry Seidel that the Sendlers endorsed the escape, but now the Bernauer tunnel crew heard conflicting stories. Uli Pfeifer told Joachim Neumann, "We will never try anything like that again with crazy guys like Fritz Wagner, especially coming up inside a house. I don't know if they really thought the couple would allow the escape or if it was some joke by Wagner!" The episode left Neumann even more worried about bringing his girlfriend through the Bernauer tunnel a few weeks ahead. Hasso Herschel, for his part, held Seidel personally responsible for not alerting them to the Sendler uncertainty.

Back at work, the diggers shored up their tunnel with wooden supports, made sure the lights and phones were still operable, and got the electric rail system running again. Piers Anderton and the Dehmel brothers returned to catch up with the renewed operation. A huge boulder had been encountered in the clay at the end of the tunnel, too mammoth to remove, so the diggers had to take a left turn around it, and back to the right, and then borrow surveying equipment again to make sure they were still on track.

Then another tunneler found out about the NBC filming. Herschel had grown suspicious about the arbitrary scheduling of shifts, and finally confronted Gigi and Mimmo. The Italians claimed they had been planning to tell him about the NBC deal anyway. Anderton had insisted NBC needed to film more diggers—no one would believe that this tiny crew had dug such a lengthy tunnel. Hasso demanded money if he was going to play a key role on camera. Coming from the East, after several years in a prison work camp, he had had no idea that media would pay even a penny for photo or film rights, but now that he knew that he wanted to exploit it. As they had done with Joachim Rudolph, the Italians gave him 1000 DM out front with the promise of 1000 more if and when the project was completed successfully.

Sesta, meanwhile, visited Peter Schmidt again in the East and assured him, after the Kiefholz catastrophe, that the original escape passage was back on track. Herschel got the same message to his sister. Although both families had just escaped arrest by the narrowest of margins, they were ready to try again in a few weeks—they were that desperate to flee. After more than a month lost to the flood, the new target date for a breakthrough over on Rheinsberger Strasse was set for October 1.

WHILE the subjects of their film started digging again, NBC News faced a new threat. Robert Manning, the State Department's public affairs director, had asked Bill McAndrew, the network's news chief, if he would meet with his deputy "about a Berlin matter that should not be discussed on the telephone." McAndrew no doubt guessed what this was about—he was the executive who had signed off on Piers Anderton's tunnel back in June. He also knew that Anderton had worked with Pierre Salinger for several years at the *San Francisco Chronicle*, and that they were still friendly. Then there was this: Anderton and Bobby Kennedy, in a wild coincidence, had attended the same Catholic boarding school in Portsmouth, Rhode Island, just a few years apart. It wasn't much, but McAndrew knew personal relationships counted with this administration.

The meeting, which had been cleared with the White House, was held in New York City. Manning's deputy, James Greenfield, told McAndrew that State had persuaded CBS to cancel their coverage of the Kiefholz tunnel after discovering through a "double agent" (no elaboration) that the mission had been compromised. Subsequently, State learned that Anderton had been spotted near the scene of the tunnel entrance on August 7. McAndrew conceded that Anderton had been there, but this was the only time he had filmed anything related to the Kiefholz project. Even then he was covering it purely "as news." NBC had, in fact, shown great restraint in not airing any of the dramatic footage. That might be so, Greenfield said, but what about the future? Secretary Rusk was extremely concerned about the involvement of TV networks with *any* Berlin tunnel, fearing this would only exacerbate U.S.–Soviet tensions.

The NBC exec decided not to volunteer that his network had already shot 6000 feet, or three hours' worth, of film in just such a tunnel. He believed (though didn't verbalize) that there were significant differences between the NBC and CBS projects. The Kiefholz tunnel had been dug in an open field. Its organizer was an unsavory character, a certain *Dicke*, allegedly out to make a buck. Security had been awful—it had become known to the

State Department, to the West and East Germans, and to NBC. In contrast, the Bernauer tunnel was an idealistic effort, and a model of expert engineering. Its security measures had been put to the test for months without incident.

Given the opportunity to come clean about the NBC tunnel, McAndrew remained mute. He did volunteer, however, that NBC had a source who claimed that, contrary to CBS's assurances, Schorr did have a cameraman at the tunnel entrance on escape day. When the meeting broke up, Greenfield felt it had gone badly. The NBC exec seemed antagonistic—a far cry from the attitude of CBS's Blair Clark, the JFK pal. McAndrew, for his part, had lost nothing and gained this: confirmation that the State Department knew nothing about the Bernauer tunnel, so far.

IT was just another evening shift in the Bernauer tunnel, now deep into East Berlin under the death strip. Four young men were doing the dig, deposit, and dump routine, as mounds of damp clay rose in the corner of the factory basement. Suddenly the lights in one area of the tunnel flickered. The winch stopped working briefly. Then, after a pause, it happened again, and a third time.

Joachim Rudolph happened to be on duty, and he walked to the head of the tunnel to investigate. He had rigged up their lighting via the factory's fuse box, which was mounted next to the old wooden door that led from the cellar to their work site. The door was usually closed and "locked" by a lonely log propped against it. When Rudolph arrived, he saw an arm squeezing through a narrow space on one side of the door, reaching for the fuse box.

Rudolph, alarmed, shouted through the door—what the hell was happening? "Open up, I know what's going on here!" the intruder exclaimed. He guessed they were digging a tunnel—he had tried a couple such projects himself that failed, he explained—and he desperately wanted to get his wife and two children out of the East. Could he join them for the duration of their dig? It was a wild tale, but Rudolph opened the door for a better look at the visitor. He turned out to be an anxious young man of average size named (he claimed) Claus Stürmer. As many as thirty students worked part-time on the tunnel, some not terribly well-known to the organizers, but they always arrived with a firm invite and at a set time. This was the first true security breach.

The diggers looked like they had seen a ghost, Stürmer observed. They were filthy and seemingly disorganized—Stürmer wondered if *he* could trust *them*—as they debated whether to order the young man to slip away and forget about the tunnel. But that would mean that someone who knew what they were doing would be at liberty in Berlin, probably resentful and talking to god-knows-who. It would be better to find out if this guy was legit and try to control him in some way, or at least keep half an eye on him. Rudolph instructed Stürmer to come back the next evening when the three tunnel organizers would be present. When Stürmer heard two of them identified as Hasso and Mimmo, he thought, *What? These are dogs' names.*

Arriving at the cellar at the appointed hour, Stürmer was surrounded by the two Italians, Wolf Schroedter, Hasso Herschel and Joachim Rudolph, who ordered him to take a seat. One of them displayed a pistol. Stürmer squirmed, feeling acutely a nail in the chair poking his ass. Another digger, suspecting Stürmer was Stasi, had already contacted the LfV to start a security check.

Stürmer, who said he was twenty-six, had quite a story: far from unique in its elements but hard to swallow whole. A butcher in the East, he had tried to flee with his wife, Inge, and six-year-old daughter the year before. He had made it through the wire when shots were fired. His wife froze; she was captured with their child and sent to prison. Pregnant at the time, Inge was mercifully released after the birth of the baby, in March 1962, and reunited with her daughter. Stürmer started digging tunnels to rescue them. The first was under Heidelberger Strasse, nearly parallel to Harry Seidel's tunnel. He gave that up after Heinz Jercha was shot. (When Inge heard that a butcher in his midtwenties had been slain she was certain it was Claus.) Then he started one along Bernauer Strasse, but his work crew quit. Desperate, he nosed around. What if, as happened at Heidelberger, others were digging nearby?

Sure enough, the previous night, he had heard sounds emanating from the nether regions of the swizzle stick factory. When he approached, he saw clay caked on the courtyard steps leading to the basement. Now he just wanted to join their team and retrieve his wife and two children, one of whom he had never seen. He would even donate his car, if necessary, and quit his job to dig full-time.

Far from convinced, the tunnelers kept asking, loudly, if he was Stasi, demanding proof of sincerity. One told him, "*You* wouldn't believe this if you were us." Mimmo Sesta was especially on edge. They also wanted to make sure Stürmer didn't possess any weapons. To find out, they demanded the key to his apartment out in the Spandau district. Stürmer told them, frankly, yes, he did have a pistol, and that they could remove it from his closet. A search by Hasso and Gigi failed to find the gun. What else was Stürmer lying about? So they tied his hands and feet and bound him to a chair. Mimmo stood behind him, pantomiming the act of pushing him in a hole. Someone shouted, "If we find out you are Stasi, you are not getting out of here!"

Stürmer provided details on exactly where the gun was hidden in the closet. On a second visit they found it, a 9 mm pistol they were happy to add to their tiny arsenal. But they also found something else—a box of letters from Inge, the alleged wife in the East, which also mentioned his two children. At least that part of the tale now rang true.

Still, suspicions about Stürmer remained, so just before midnight they threw a blanket over him—still bound with ropes—and hustled him to the VW van. As he was driven to Tempelhof Airport to meet LfV interrogators, one of the diggers warned,

"If you try to escape we will shoot!" Stürmer felt like he was in a Hollywood gangster movie. Told where he was going, Stürmer thought, *Whoa, they really have connections—maybe they are going to fly me out of Berlin so I won't betray them.* He had dreamed about such a flight, though not under these circumstances. Alas, he wasn't going anywhere beyond Tempelhof. The LfV grilled him all night. One of the intel agents showed Stürmer the box of letters from his apartment. "Love letters," Stürmer explained. "Sometimes two a day!" They were unable to shake the rest of his story and so, that morning, they released him, telling the tunnelers that he might be clean after all.

After that nightmarish experience, would Stürmer count his blessings and search for another way to free his family? Not a chance. Hours after his release from the LfV, he showed up at the Bernauer tunnel again, reporting for work. "I do not hold a grudge," he insisted.

The tunnel organizers huddled again. What to do? The only way to keep an eye on him was to make him this offer: take a shift digging every single day. In exchange they would allow passage for his wife and kids—but they would be the last refugees to be notified on escape night. Friends and family members of the other diggers would have to make it through the tunnel safely, and only then would Stürmer be allowed to contact his wife. If she wasn't home when that happened, tough luck. The tunnelers reminded Stürmer that they had weapons and weren't afraid to use them. Or they might imprison him in one of the dungeon-like rooms in the cellar until the end of their project if suspicious new information about him surfaced. In case he really was Stasi, they vowed to never tell him even vaguely where or when they planned to break through in the East. Stürmer accepted all this with alacrity and picked up a shovel.

JUST days after it was proposed, the MfS awarded the People's Army Silver Medal for Merit to the fast-rising Stasi informer "Hardy." Siegfried Uhse would also get a cash payment of

1000 DM (or $250)—no small change given his regular MfS
pay of 100 DM about twice a month. Considering what Uhse
accomplished with the Kiefholz tunnel bust, one wonders what
the Stasi reserved its gold medal for. His handlers wrote a two-
page tribute:

> Through reports and information by the IM [informant],
> a large-scale, organized and forceful breakthrough by a
> West Berliner terrorist organization at the border has been
> prevented and the involved persons have been arrested.
> Armed bandits planned to penetrate into the state territory
> of the GDR by means of a tunnel from West Berlin to the
> Democratic Berlin and assume the securing of the border
> breakthrough by force of arms.... [T]he forceful border
> breakthrough was to be prominently exposed by the West-
> ern press, radio and television....
>
> The IM proved operational readiness and commitment,
> reliability, pro-active/self-initiative and personal courage.
> By cautious performance of the IM, it was possible to ar-
> rest a total of 40 individuals, including four members of the
> West Berliner terrorist organization that substantially par-
> ticipated in the organization and execution of the provoca-
> tions at the border of the Democratic Berlin.

The award was approved in a handwritten notation on the
front page: *Agreed, Mielke.* The Stasi boss of bosses, Erich Mielke.

BY now, the Stasi had completed another internal report on the
Kiefholz bust, listing the arrests so far at forty-six. Six of the
prisoners were to be kept under tight control of the MfS, including Wolf-Dieter Sternheimer, the Stachowitzes, and the Sendlers. The report included an unusual section titled "Motives
to Illegally Leave" the GDR, based on the interrogations. The
result: eleven people had wanted to join girlfriends or boy-
friends in the West, with the same number citing relatives or

LEFT: Harry Seidel, an East German cycling hero, fled to the West before freeing dozens of others via tunnels and other methods of breaching the Wall.

RIGHT: Official Stasi ID photo for informer Siegfried Uhse, code named "Hardy," whose actions in 1962 would lead to the arrest of many dozens of East Germans.

ABOVE: Divided down its center by the Wall, Heidelberger Strasse became known as the "Street of Tears"—and as the site of a remarkable number of early tunnels.

ABOVE: Joachim Neumann, engineer, aimed to secure a way out of the East for his girlfriend.

ABOVE: The three organizers of the 1962 tunnel beneath Bernauer Strasse—subject of NBC's *The Tunnel*—on the streets of Berlin: (*from left*) Gigi Spina, Mimmo Sesta, and Wolf Schroedter.

ABOVE: Joachim Rudolph served as chief electrician on the Bernauer tunnel and took part in breakthroughs in the East.

ABOVE: Hasso Herschel vowed not to shave his beard until his sister, Anita, and her daughter came through his tunnel safely. He led two breakthrough operations to achieve this.

LEFT: Key NBC figures near the site of the tunnel on Bernauer Strasse with a backdrop of boarded-up buildings: *(from left)* Berlin bureau chief Gary Stindt, producer Reuven Frank, and correspondent Piers Anderton. Harry Thoess is likely behind the camera.

BELOW: NBC correspondent Piers Anderton at the entrance to the Bernauer tunnel, which started with a triangular frame before being converted to a square design.

LEFT: A serious water leak in early summer 1962 brought tunnel building under Bernauer Strasse to a halt.

LEFT: Daniel Schorr of CBS found his Berlin tunnel much later than did Piers Anderton, but was nearly the first to produce a film—until the State Department, the White House, and his boss intervened.

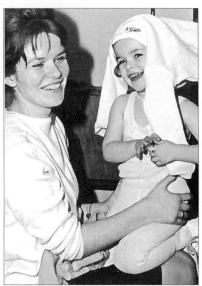

RIGHT: Rescuing the Schmidt family (Eveline Schmidt and her daughter, Annett, pictured here) was the original impetus for the Bernauer tunnel.

ABOVE: Secretary of State Dean Rusk, with the support of President Kennedy, tried to halt network television coverage of the tunnels.

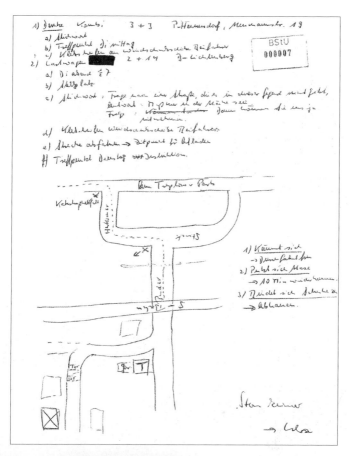

ABOVE: Sketch of the area around the home of Friedrich and Edith Sendler, in whose living room the Kiefholz Strasse tunnel would break through. On the morning of the escape, a courier gave the sketch, marked with drop-off points for refugees and signals they were to look for, to Siegfried Uhse.

LEFT: August 7, 1962: Stasi agents enter the Sendler home to apprehend the Kiefholz tunnelers.

ABOVE: August 17, 1962: Iconic photograph of East German border guards recovering the body of Peter Fechter, age eighteen, shot and killed in the act of trying to escape to the West.

ABOVE: The view across the
Wall to Schönholzer Strasse
from the apartment NBC
rented. Number 7, where
the Bernauer tunnel would
surface, is at far right.
A sheet hanging in the
apartment's window signaled
couriers and escapees that the
coast was clear.

LEFT: September 14, 1962:
Ellen Schau, a courier, was
filmed by NBC's Peter Dehmel
as she approached an S-Bahn
elevated station for her fateful
trip to the East.

LEFT: Eveline Schmidt,
carrying her pocketbook,
was first out of the tunnel.

RIGHT: Anita Moeller, sister of Hasso Herschel, in her torn Dior wedding dress after crawling through the tunnel.

LEFT: Harry Seidel's final tunnel, completed in November 1962, followed this path from the West to a target house in the East. The Stasi planted explosives between the two houses.

LEFT: Four of the key Bernauer tunnelers, almost a half century later, outside the newly renovated 7 Schönholzer Strasse: (*from left*) Uli Pfeifer, Joachim Rudolph, Joachim Neumann, and Hasso Herschel.

friends; five said they had expected better living conditions across the border; two had wished to avoid the military draft; one had simply wanted to go to school in the West.

Couriers continued to undergo lengthy interrogations, day and night. After a few such sessions, Sternheimer had concluded that a source within the Girrmann Group must have given the escape away, and if so, the snitch was likely that thin, young man in the nylon jacket, Siegfried Uhse. It seemed strange to Sternheimer that Stasi interrogators often asked him about fellow couriers, such as Hartmut Stachowitz, and other Girrmann associates, but never inquired about Uhse. Also, they seemed to know details about Sternheimer's apartment, and only Uhse among the Girrmann team had ever stepped inside there. Unfortunately for the Girrmann Group, Sternheimer had no way— and might never have a way—to pass along his suspicions. For those interned in a Stasi prison the outside world ceased to exist, indefinitely.

Manfred Meier had once believed the Stasi might not hold him for long because, as far as he knew, neither his girlfriend Britta nor fellow plotter Sternheimer had been arrested. Then he found out they had seized Sternheimer—and an interrogator told him, "Best regards from Britta Bayer."

Also undergoing heavy questioning was Fritz Wagner's main courier, Dieter Gengelbach. An eighth-grade dropout who had fled to the West in 1956, married, and worked in butcher shops, he was arrested eight days after the Kiefholz fiasco. Despite the danger that a detainee would rat on him (which indeed happened), he had foolishly revisited the East. Gengelbach quickly admitted to the Stasi, "I am a member of a smuggling organization."

Gengelbach was just as direct when asked about the genesis of the Kiefholz tunnel. He revealed that the famous cyclist Harry Seidel had picked the spot the previous winter, found the water table too high, but returned in June, beginning in "a mini-bunker surrounded by bushes" and aiming for the Sendler cottage. After digging the first two-thirds of the passage, Harry had notified the LfV man Mertens that he needed help. Mertens contacted another team who finished the excavation and the "preparations for the smuggling." Someone named Hasso had been in charge of "the student group" doing the final digging.

Asked about other tunnels, Gengelbach offered: "I also know that the student group [the one led by Hasso] is building a tunnel from a property somewhere in Bernauer Strasse. This tunnel is supposed to be 150 meters long, seven meters deep and the work is difficult due to hard clay soil. I also heard that near this tunnel is a water pipe broken on the territory of East Berlin." This was at least the third tip the Stasi had received that month about the NBC tunnel.

Friedrich Sendler had also endured repeated interrogations, and changed his story over time. First he admitted discussing a possible tunnel with a visitor as far back as February (around the time Harry Seidel briefly considered tunneling in the area). A foreman with the electric company had observed that the Sendler cottage was perfectly situated close to the border for a tunnel breakthrough. He dreamed of joining his mother in West Berlin—perhaps that could be arranged with Sendler's help? Sendler claimed he took his remarks merely "as a random idea."

Then one or both of the Sendlers disclosed mysterious phone calls or visitors to their home in the days before the Kiefholz bust. It seemed that on August 4, three days before the escape attempt, a certain Gengelbach had called Friedrich Sendler to find out if he was hiring any helpers in his workshop. Nothing supposedly came of that. (What a coincidence: Fritz Wagner's courier calling about a nonexistent job.) Then, on the eve of the escape, a stranger appeared on the Sendler doorstep asking about a carpentry job, but only Herr Sendler's mother had been at the house. That man, or another—perhaps desperately trying to confirm that the Sendlers were on board with the escape— even called the next day, with refugees about to make their way to the site, and asked if the nonexistent "job" was still available.

The Sendlers repeatedly denied offering encouragement to escape helpers, but the Stasi still wasn't buying it. An MfS report concluded that when Edith Sendler left her home on August 7 after noticing the drilling in her living room, she and her husband had "wanted to leave their property." When the Stasi in their front yard made it clear that wouldn't happen, Frau Sendler "couldn't help" but reveal what was going on under her floorboards.

AUGUST 17 dawned gray and drizzly in Berlin. Peter Fechter and Helmut Kulbeik, the young bricklayers, toiled most of the morning at the former Kaiser Wilhelm Palace, then joined two fellow workers for a pint of beer at a pub near the restricted zone. Leaving to return to work under the dull noon sky, they realized this provided a golden opportunity to break from the group and make their escape. Perhaps they were caught up emotionally in the protests that had raged for four days since the Wall's anniversary. Peter was still upset that the GDR had turned down his request to visit his sister in the West.

Telling their colleagues they needed to purchase a pack of cigarettes, the two youths grabbed a quick bite to eat and headed for the factory they had scouted several days before. Situated directly on the border strip at Schützen Strasse, two blocks south of Checkpoint Charlie, it was within sprinting distance of the Wall. Wearing their blue work clothes, a kind of disguise, they entered the building from the street and stepped quietly. They found windows in the back room on the ground floor bricked up, save one. A small opening, crisscrossed with barbed wire, let in some light. Nearby on the floor was an enormous pile of wood shavings, where they decided to hide and, hopefully, nap, until that evening, when encroaching darkness would provide some cover. The August heat—and no doubt their anxiety—made sleep impossible. More than an hour passed in silence.

Just before 2:00 p.m., the pair heard voices in the factory. Someone may have spotted them entering the building. Fearful of being discovered, they decided to make a run for the Wall, about twelve yards away, in broad daylight as soon as the voices faded. They pried out the nails and staples along the window and pulled down the wires. Peter, a few inches shorter than his friend, squeezed through the window just as Kulbeik noticed a carpentry worker entering the room. The worker spotted them but, startled, said nothing, then left, perhaps to inform others. Kulbeik followed Fechter through the window, landing as close as possible to the side of the building to remain in the blind spot of GDR guards a few hundred feet to the right. Then they hurtled over a pair of barbed wire obstacles and, with Peter a few steps ahead, dashed across the narrow death strip toward the Wall, now only five yards away.

Without the warning required by GDR rules, shots from two Kalashnikov automatic rifles rang out. Kulbeik sprang past Fechter and managed to scramble up the eight-foot Wall, then forced his way through the wire strung on Y-shaped metal brackets at the top. He was ready to swing over, with only a few scratches to his chest, when he saw Peter still standing motionless at the foot of the Wall, probably too shocked by the sudden bursts of gunfire to move.

"Hurry up, go on with it, jump!" Helmut shouted at Peter, and dropped down the Western side of the Wall. Fechter may have considered trying to climb the Wall but, from a dead stop, sensed this might be futile. His only option: take cover behind one of the concrete supports that jutted out from the Wall every few yards, while deciding whether to try to mount the barrier or perhaps surrender. But the supports, barely a few feet wide, only shielded him from shots coming from the right. Now heavy firing started from the opposite direction as well. (Some of the shots hit the building to the right of Checkpoint Charlie in the West, the duty officer's log for the U.S. military noted at 2:12 p.m.) East Berlin border guards Erich Schreiber and Rolf Friedrich, both recently drafted into service and given little training, had aimed about two dozen shots at the helpless Fechter from their station. One 7.62 mm steel bullet hit Peter in the pelvis and exited the other side of his body. Fechter collapsed, bleeding heavily and screaming in pain, half a yard from the Wall.

Observing all of this was seventeen-year-old Renate Haase,

who was waiting for her boyfriend outside a nearby office. Having recently completed a Red Cross course, she felt obligated to help, but when she tried to run to the wounded youth a border guard shoved her back, shouting, "You can't go there! Shots are being fired!"

She yelled back at the guards, "You swine! You criminals!" and demanded that someone take photos of the shooters. Onlookers standing near her pulled her back.

Meanwhile, across the Wall, a dazed Kulbeik ran toward the headquarters for press baron Axel Springer. He spotted two U.S. Army soldiers and called to them. The soldiers led him to a Jeep and drove away. Then a West Berlin police patrol car arrived. Knots of citizens had gathered there on a wide strip of open ground, from which they were clearly able to hear Fechter's screams of *"Helft mir doch, helft mir doch!"* ("Help me, why aren't you helping me?") A cop shimmied up the Wall and stuck his head through the barbed wire crown, but West Berlin police were instructed not to cross the demarcation line under any circumstances. He saw the young man lying on his back beside the Wall. Another police officer tried to talk to Fechter and threw bandages over the Wall, but the victim, too weak to reach them, rolled onto his side in a fetal position.

At 2:17 p.m. a young American lieutenant at Checkpoint Charlie dialed Major General Albert Watson, the commandant of Berlin's American garrison, asking for instructions. Watson told him: "Stand fast. Send a patrol but stay on our side!" As six MPs arrived at the scene, a crowd of about 250 West Berliners were shouting into the East, "You criminals! You murderers!" Bypassing the normal chain of command, General Watson called his commander in chief directly at the White House, asking for orders from Kennedy on what to do next. JFK was visiting Colorado, but his top military aide, General Chester V. Clifton, conveyed the situation to him. "Mr. President," he said, "an escapee is bleeding to death at the Berlin Wall."

By 2:40, half an hour after the first shots, a German interpreter for the U.S. military reported that one youth was "wounded and is lying against the wall on the east side. He is able to talk to the personnel on the west side of the wall." The Americans kept an eye out for any movement of Eastern troops toward the shooting scene. The West Berlin police requested a U.S. ambulance.

One of those who heard Fechter's cries was Margit Hosseini, a bookseller's apprentice who had been visiting friends near the Wall in the East. They heard something happening at the Wall, and went out to look from a fourth-floor window. She observed a male body lying in an S shape and heard the victim scream or cry for help, before his voice grew weaker and weaker, until it stopped. It left her feeling useless, traumatized.

A young West Berlin newspaper photographer, Wolfgang Bera, had also heard the shots and ran to the scene. At first he thought there was nothing to see, but then he noticed an elderly woman in a window of an apartment building across the Wall. She pointed a finger at the Wall near him, then drew the curtain shut. He understood immediately. Bera found a ladder, climbed up on the Wall, and looked down to see the bleeding youth just below him. Pushing his small Leica through the barbed wire, he captured Fechter, on his side, with his right hand stretched out and his hand open. Blood was trickling into his palm, filling it. Bera climbed down and, figuring the paralysis shown by police on both sides meant that only the Americans were in a position to take charge, ran to Checkpoint Charlie and asked for help. The soldiers appeared indifferent. One GI told him: "That's not our problem." The four words would soon become infamous.

Also in the neighborhood was freelance journalist Herbert Ernst, who routinely searched out newsworthy events to film in West Berlin—often driving his VW Beetle hundreds of miles a day along the sector boundaries. He'd just stopped to buy a new lens at a store on Friedrich Strasse when he heard the shots. Soon he was among the gathering crowd of West Berliners, with soldiers milling about, as a U.S. helicopter circled overhead. He mounted one of the viewing platforms scattered along the Wall in the West and immediately began filming it all, though he didn't yet know what was going on. As word spread, more newspaper reporters and photographers arrived. East German guards responded by hurling tear gas in their direction, attempting to curtail damaging news coverage. They also fired tear gas near Fechter, perhaps to hide the body from cameras and onlookers or to enable them to retrieve him under cover of fog.

Returning to the Wall, Bera attached a telephoto lens to his Leica, stood on a low pedestal stuck in the ground and, lifting his camera high over his head, captured what would become one of the iconic photographs of the Cold War era: four East German border guards finally hauling off Fechter's inert body fifty minutes after he'd been shot. Another arresting image: one of those guards carrying Fechter in his arms toward a police vehicle for transport to a hospital.

Mounting a small podium, Ernst captured the same moments on film. His camera recorded a VoPo gathering Fechter under the armpits and another taking him by the feet—*like a wet sack*, it seemed to Ernst—as they hurried down Charlotten Strasse. (This sequence, lasting just forty-four seconds, provided the first footage ever of an escapee's body being retrieved at the Wall.) As they did, they passed Renate Haase and her boyfriend, who cursed the VoPos loudly. The young couple, who were planning to announce their wedding engagement the following day, were promptly arrested and transported to a nearby police station.

General Watson again called White House officials—who had sent no orders—and informed them, "The matter has taken care of itself."

BY late afternoon hundreds of angry West Berliners had gathered near Checkpoint Charlie. At 5:00 p.m. the First Border Brigade in East Berlin was informed that "the injured was now deceased." Almost two hours later, two young East Berliners holding a small sign appeared in a window on the fourth floor of a building near Checkpoint Charlie. An American interpreter across the Wall managed to signal them to use a larger sign, which they did.

It read, *He is dead.*

As the evening wore on in the West, the protest crowd grew into the thousands. Many shouted, "Murderers!" over the Wall at three East Berlin guards who stared back, impassively, automatic pistols at the ready. Other VoPos threw tear gas grenades across the Wall. Police in the West retaliated with their own canisters.

Then the rioting erupted. Crowds began throwing stones, bottles, and bits of iron bars at the East German border police. VoPos responded with smoke bombs and tear gas. West Berlin authorities sent police to disperse the crowd, but the crowd turned on them. When Allied military police drove up in Jeeps, the crowd, incensed that the Americans had not tried to help Fechter, shouted, "Yankee cowards! Traitors! Yankees go home!" Into this maelstrom a bus that took Soviet sentries to guard their war memorial in the British sector tried to make its usual way; the crowd hurled stones and broke the bus windows. Russian soldiers could be seen cowering inside, shielding their faces with their arms, but the bus didn't slow. When it returned, it needed an escort of British military police to shepherd it to the East.

By then, Wolfgang Bera's dramatic pictures were being readied for publication in the *Bild Zeitung*, while Herbert Ernst's footage had been purchased by a leading West German TV station for 100 DM (plus two bottles of whiskey from a photo agency that acquired stills from the film).

In East Berlin a young photographer named Dieter Breitenborn, who worked at the magazine *Neue Zeit* on Zimmer Strasse, was about to develop his own images. He had captured another perspective on the tragedy that afternoon from a window in his office high above the Wall: a boy's body at the base of the Wall, the circling helicopters, the smoke bombs, then the guards' belated retrieval of the lifeless victim (captured in a sequence of disturbing frames). As Breitenborn began his work in the darkroom, he heard a knock at the door. A coworker, long suspected of being a Stasi informer, demanded, "Give me the film!" Breitenborn felt helpless to resist.

Late that evening, other Stasi agents arrived at the Fechter family's apartment in Weissensee. They demanded to know where Peter Fechter was and searched the entire flat for weapons or incriminating political literature, but came up empty. Finally they hinted to the Fechters, but refused to state directly, that their son and brother might have been shot at the Wall that afternoon.

11
The Martyr

The murder of Peter Fechter hit West Germans like a punch in the heart. The front page of the country's largest newspaper, *Bild Zeitung*, featured an enormous version of Wolfgang Bera's photo of four guards hauling Fechter away as a headline declared, "VoPos Let 18-Year-Old Bleed to Death—As Americans Watch." The *Morgenpost* published the same photo under Fechter's dying plea in huge letters, *"Helft mir doch, helft mir doch!"*

The story also rippled to distant shores. The *Guardian* in London observed that police had carried Fechter away "like a sack of potatoes." The *New York Times* published on its front page a different Bera photo—two guards carrying the lifeless, unnamed victim to a car—under the heading: "German Reds Shoot Fleeing Youth, Let Him Die at the Wall." East Germany's *Neues Deutschland*, on the other hand, reported that two "fugitive criminals," supported by West Berlin police, had tried to flee. Since they had not responded to "repeated requests and warnings" (a false claim), the border patrol had been forced to resort to "use of firearms." An official morning-after autopsy, not made public, found that only one bullet of the nearly three dozen fired had struck Fechter, through the hip. Had some of the East German guards tried to miss?

Violent demonstrations broke out near the site of the kill-

ing, outside American military compounds, and at Checkpoint
Charlie, where protesters held a banner directed at the United
States reading: *Protecting Power = Murder Helper.* The daily bus
that transported Soviet soldiers to their war memorial in the
West was once again stoned and a window broken. American
officials in Berlin allowed that "indignation" in the streets was
understandable, and promised to seek new ways to respond to
brutality at the Wall. A U.S. military officer told Arthur Day,
chief of political affairs at the Berlin Mission, "This is going to
have serious repercussions."

A large wooden cross and a wreath were erected in the West
yards from where Fechter fell. Across the Wall, East German
police obsessively monitored this memorial site. At 8:40 that
morning they counted fifteen people leaving flowers there; at
10:25 another hundred. At noon more than five hundred protesters had gathered. Fechter's sister Liselotte arrived with her husband to lay another wreath. A GDR border guard peered over
the Wall and jotted down the inscription: *In silent remembrance,*
your sister Lilo, your brother-in-law Horst. You wanted freedom
and had to die.

Improbably, on this day, a teenage girl crawled under the
wire in the Spandau district and drew heavy fire, but escaped
to the West. Not wounded, she nevertheless went into shock and
was hospitalized.

The young men digging from Bernauer Strasse deep under
the death strip were also furious about the Fechter shooting,
which had occurred just three miles to the south. Fechter, like
many of them in the past year, was determined to escape the
East, although his attempt had been especially risky—and
failed. Some of the tunnelers who had secured a deadly weapon
said they would be more likely to use it now, if necessary, in an
escape action. Others confessed to wanting to rush out right then
and shoot a VoPo in revenge. One of the tunnel organizers cut
out graphic newspaper photos of Fechter, bleeding at the Wall or
carried off dead, and posted them at their work site, for inspira-
tion. Joachim Neumann reflected, *It's just like the East German*

regime—to not only shoot a boy, but then refuse to help him. When muscles or spirits sagged, pictures of Fechter—the hero, the martyr—strengthened their spirit.

The official East German report on the Fechter shooting quickly concluded that the guards' actions were "correct, effective and determined. The use of weapons was justified." Two of the shooters and two officers at the scene received awards. Daniel Schorr, sent by CBS to cover a fizzling story on Cyprus—and maybe get him out of the State Department's line of fire—felt angry about being so far from the latest drama, and petitioned for a quick return.

HARTMUT and Gerda Stachowitz still did not know the whereabouts of each other, nor of their infant son, Jörg, taken from them after they were seized in the Kiefholz bust. Gerda remained consigned to the women's prison on Barnim Strasse (where Marxist firebrand Rosa Luxemburg once dwelled), with no trial date in sight. Hartmut, at the dreaded Hohenschönhausen, prepared for what he knew would be a public show trial at the end of August. Each had been harshly interrogated, day and night. Hartmut, the veterinarian, was accused of squandering the fine education the Communist state had given him.

Unlike many others arrested, he refused to provide the Stasi with fresh names or details about the smuggling operation, frequently responding to questions with, "I don't know." (Siegfried Uhse's chief handler, Herr Lehmann, was in charge of some of the interrogations.) He played dumb when asked to identify Sternheimer, although he did give up the name of Joan Glenn, knowing she was safely in the West. When he talked about meeting Uhse, he simply called him "the man in the nylon jacket," adding that he appeared to be in good standing with the Girrmann Group.

Whenever Gerda asked about her son, a Stasi interrogator would berate her, "You are not fit to raise a child in a socialist society," or "You might see him again if you tell us everything you know." Hartmut regretted that he never had a conversation with his wife about what to say under these circumstances, which neither had dreamed would come to pass. Both soon learned that while it was wise to reveal at least a few true details when interrogated, it was imperative to say nothing in the general prison population. The Stasi, no less than in the outside world, had a system of snitches in place, and if you complained about your treatment or prison conditions you were likely to have another charge added to what you faced—*staatshetze*, or rabble-rousing.

One turn of events might have comforted them. Gerda had last seen Jörg, asleep on a table in a soaking wet diaper, in an interrogation room on the night of her arrest. The next day, unbeknownst to her, he had been placed with an East Berlin family, who had promptly enrolled him in a day care center. Later that week, a staffer there recognized his last name: The daughter of a woman she knew had become a Stachowitz after getting married. That woman turned out to be Jörg's grandmother, who had not seen the boy since Gerda left with her husband for that "coffee with friends" on August 7. Now the staffer delivered the boy to his grandmother, a heartwarming reunion—if one could forget that his parents were stuck in a police state prison, indefinitely.

MANFRED Meier, also imprisoned at Hohenschönhausen, continued to be questioned by the Stasi, sometimes for hours at a stretch. They badly wanted him to admit that West German commandos had planned to shoot up Kiefholz Strasse on escape day. He said he knew nothing about that. Nevertheless, he was about to become a TV star. One day a Stasi agent informed him, "Good news! You may be here wrongfully. You have a chance to defend yourself!" The Stasi planned to put him on state-controlled TV where "you can tell your story."

The MfS was leaving nothing to chance. On August 20, the day before the broadcast, a Stasi staffer in the wonderfully named Department of Agitation/Propaganda composed a detailed scenario—in fact a partial script—for the show. "Target of TV talk should be to prove" that on August 7 a "violent border provocation" was prevented only by "the intervention of security organs of the GDR," the scriptwriter urged. The program would open with the East German commentator declaring that the Girrmann Group was behind this tunnel and used the "firearms and terroristic" tactics they inevitably favored. The TV host should display maps and photos of the crime scene. Then Meier, one of the tunnel's "organizers," would be interviewed about his meetings with Sternheimer to discuss the operation and his "reconnaissance" of the Sendler property. Meier would not be asked about any weapons, since it appeared likely he would deny it— the last thing the Stasi wanted to happen.

The script continued with what Edith Sendler was to say in explaining what had happened on escape day. She was to express "indignation" over this uninvited invasion of her home, after which the commentator would display photos of the gaping hole in her living room floor. He would hail the Stasi agents for halting the invaders, leaving unsaid that they then let them all escape. After that, in the script, two Westerners who had allegedly worked with the Girrmann Group were to testify about the purchase of American machine guns and possible use of explosives.

The next morning Meier was given his civilian clothes. Three Stasi agents blindfolded him and marched him to a limousine with darkened windows for the drive to the TV station. (Before covering Meier's eyes, one of the Stasi men hoisted up his jacket and displayed a pistol, saying, "Just so you don't get any stupid ideas.") When they got to the studio, Meier declined a cup of coffee, fearing he might be drugged or poisoned. "You can drink the coffee," an agent assured him. "Only real Cuban beans!"

Then he was interviewed by the GDR's chief press officer as well as famous radio propagandist Karl-Eduard von Schnitzler. This was a taping for airing later that day, as the Stasi would never trust the uncertainty of a live broadcast. Meier admitted he had taken part in aiding refugees (he could hardly do otherwise), but that wasn't enough. Over and over the program's hosts tried to get Meier to admit that escape helpers were heavily armed on August 7 and had planned to spark a bloodbath. He denied it, saying that he would never be a party to violence, and that he had seen no weapons that day. When the interview was over, Meier reflected that, since the session was taped, it would no doubt be edited to twist his answers.

That's just what happened. His fear of a bloodbath was edited to suggest this was because he knew the West German "ultras" and "gangsters" planned to initiate one. The next day the East Berlin newspaper *Neues Deutschland* covered the interview, with a large photo of Meier ("member of the notorious terrorist Girrmann Group"), wearing horn-rimmed black glasses, at the studio. The headline declared, "Instigators in Bonn and USA Prepared Bloody Actions and Murder."

RIOTING in West Berlin over the Fechter outrage—against the Soviets, the Americans, the West Berlin police, maybe the entire human condition—continued with no sign of ending soon. Another Soviet bus was attacked and more windows broken. A mob of several thousand people, mainly young, broke through cordons, threw rocks over the Wall, and tried to set two cars on fire. Some held placards over the Wall for guards to see, warning the *Murderers* that their *Day of Reckoning* was coming. The anti-Soviet angle, at least, did not displease the U.S. Mission. Its political chief, Arthur Day, explained in a memo to a superior that "we in the Mission and General Watson" agreed that "we should keep the initiative with the Soviets" that resulted from the Fechter incident. The Soviet commandant in Berlin, however, had already rejected the official U.S. note protesting the killing and demanding future restraint.

More troubling for the State Department were the protests' anti-American overtones. The comment by the U.S. soldier at the scene of the Fechter murder—usually reported as "That's not our problem"—had provoked wide resentment. "I don't want to exaggerate this sentiment," Day wrote, "but it does exist and it is difficult to say right now whether it will die away or increase."

This prompted a peeved Dean Rusk to cable the Berlin Mission, ordering them to not "shrink from use US forces if this becomes necessary to maintain law and order . . . if West Berlin police prove inadequate." West Berliners could not have it both ways—"they cannot have our protection on basis occupation rights and on other hand ignore our directives." What he feared was a Soviet move into West Berlin to put down civil disorder. West Berlin police, perhaps in response to the Rusk warning, finally took out their truncheons and trained water cannons on the crowds until the liquid was exhausted. Protesters threw rocks and firecrackers at the police, chanting, *"Mauer muss weg"* ("The Wall must go"). Seven police and twelve civilians were injured in what the Associated Press called the worst night of protests since the Wall went up. Daniel Schorr, now back in Berlin, reported, "Peter Fechter's body has become a symbol of East-West conflict as John Brown's body came to symbolize the North-South conflict" before the Civil War in America.

On August 20, General Clifton, Kennedy's military aide, sent a rather colloquial memo to diplomatic and military officials: "The President wanted to know the facts on the man who was killed and left lying there. What rules were followed to let us not touch the guy?" The President, he added, "would like to know what gives." Mayor Willy Brandt proposed that the Allies provide an ambulance always on call near Checkpoint Charlie—and demand entry into the East to treat anyone shot at the Wall. When the ambulance got the go-ahead from the Allies, Charles Hulick at the Berlin Mission cabled Rusk that the commandants hoped this might move East Germans to take action to handle problems themselves.

THE NBC-backed excavation, in its second week since work resumed, was going well. Claus Stürmer had bolstered the effort by taking double shifts, sleeping at the site many nights on a battered couch. Stürmer felt he was better suited for this job than his comrades—he was used to physical labor while most of the students were not. He also hoped his steady presence would ease suspicions about his Stasi connections. (Stürmer remained amazed that he had found this tunnel while the Stasi, despite all their informers and resources, apparently could not.) Most of the other diggers still refused to talk to him, and he felt some might be happy if he suffered a work-site accident. The silence was harder to endure than the daily physical toll. He was also haunted by a new fear: even after all his work, *What if they don't notify my wife about the escape?*

The workers suffered a scare when Wolf Schroedter, who had been bothered by mysterious internal pain for some time, nearly collapsed one night and was rushed to the hospital. The diagnosis was kidney stones. He would recover soon, with the happy aid of drinking beer to help pass the stones, but any strong exertion—say, hacking through clay—was now out of the question. A couple of dozen colleagues were ready to take up the slack, however.

The time off allowed Schroedter to ponder something else that had been bothering him for quite a while: the NBC deal. It no longer seemed right to him for the core group to keep the filming secret from dozens of others involved. By bringing in the NBC staffers, organizers had expanded the number of people who knew about the tunnel, putting all of the diggers at risk. Then there was the matter of taking money from the network, some of which Schroedter had pocketed himself. Sure, most of it was being spent on equipment and day-to-day supplies, but the tunnel originators would likely end up making a small profit off the enterprise after they divided the extra 5000 DM from NBC at the conclusion, as well as from the sale of photos. This didn't seem right, even though it could be (and had been) easily rationalized—they had toiled longer, and thereby took greater risks, than any other diggers, not to mention sacrificing their college studies.

Still, the arrangement rankled Wolf almost as much as the kidney stones. The stones would pass, but not the rising sense of guilt.

* * *

EXCEPT for the tunnelers themselves, no one was happier about their rapid progress than Piers Anderton. NBC had expected the project to be completed by now, but at least it was back on track. Also recovering was his reputation. After his attorney threatened a libel suit, *Variety* had taken the rare step of publishing a partial retraction of its front-page story back in April on Anderton's speech to the women's club in Germany. The new piece, under the bland headline "Clarify Piers Anderton Speech," acknowledged that the correspondent had not accused NBC of "muzzling" him. And his remark that NBC wanted to put "a wall" around him was "facetious" and "misinterpreted." Another error: that he was "in hot water" with his bosses at the time.

The corrections continued. Anderton had not claimed that the American public did not care about Berlin. Actually, he had said that Americans seemed more interested in Berlin than anyone, including the West Germans. And Anderton never suggested that the United States would not stand firm defending Berlin. Indeed, he had disclosed that President Kennedy told him that America would initiate a nuclear war over Berlin if need be.

Anderton believed the original report had been orchestrated by a State Department official in an attempt to get NBC to fire or transfer him. For now, at least, his position remained secure. Confident in both the tunnel and his ability to cover it, Anderton wanted to share the renewed optimism with Reuven Frank. Their only rendezvous so far had been in London, where Anderton and his bureau chief, Gary Stindt, had flown after the major flood to discuss new fears about the project. There, over a long and lavish dinner at the Savoy Grill, Frank had decided to push ahead.

Now Anderton alerted his boss that another exotic out-of-town meeting was necessary. This time the trio met in Paris at the famed restaurant Maxim's. It was about as far as possible from the clammy confines of the tunnel, where sandwiches quickly went bad (there was no refrigerator). Anderton revealed that the breakthrough was now in sight. They talked about what their eventual film might look like. None of the footage shot so far had left Berlin, so Frank was flying blind. They envisioned that the NBC special would explore all of the escape methods tried during the past year, from sewers to early tunnels. Anderton had footage covering all of that. The Bernauer tunnel would be the key component and climax but, until it was finished, successfully or not, there was no way to judge how strongly it might carry the show. If the project failed miserably, perhaps they wouldn't include it at all, especially if the State Department found out about it.

Frank, who had not visited Berlin since the start of the tunnel, told Anderton he would not do so until the breakthrough was imminent. "When do you want me?" he asked. Anderton assured him that this time delays would be minimal.

ALL of the cables on the latest Berlin crisis did not satisfy the one man they really needed to satisfy, President Kennedy. On August 21, after returning to Washington from his long swing through the West, JFK sent a brief memo to Secretary of State Rusk: "While I recognize the difficulties the Commandants face in Berlin, I would appreciate a detailed report on the refugee incident last Friday. I would like to know how long it was before the Commandant was informed on what was happening and what action he took." He also asked what was being done to prevent the further stoning of Russian troops.

Not willing to wait for the write-up, the President phoned Rusk and asked if there was any contingency planning for the "sort of an incident" created by "this refugee [Fechter]." After all there had "been so many ones that have been close to that."

Rusk: Well, we have some for larger episodes. . . .

Kennedy: What about a single episode like this, though?

Rusk: No one case is like another. And the canals are . . .
for example, our people don't fish them out of the canals.
That's handled on the eastern side. I think perhaps the
mistake that was made locally was not offering some
medical care [to Fechter].

Kennedy: Yeah. . . . Of course, the West Berliners [protest-
ers] are not very generous but . . . that's all right. I think
we just have to ride through this one.

Later that day Kennedy met with his United Nations ambassador, Adlai Stevenson, who had just returned from meeting
with European leaders. Stevenson reported that nearly everyone
privately opposed the reunification of Germany, the Nazi era still
fresh in their minds. Kennedy wondered why none of the leaders
ever came out against a reunited Germany in public. Then he
analyzed Nikita Khrushchev's desire to unify Berlin with the
Allies removed, musing, "I don't blame him for wanting to get
us out of Berlin. . . . They have the Wall, they have this country
that's always disturbed as long as there's Berlin." But there was
one catch: "The problem is that our getting out of Berlin would
be disastrous to us."

THE CBS tunnel coverage might have been quashed, but a quite
different media outrage continued to occupy the President's attention: the leak exploited by Hanson Baldwin in the *New York
Times,* and how to discourage others like it. Kennedy had forwarded his intelligence panel's plan to monitor journalists to
CIA director McCone and others, and expected action to "protect our intelligence and our intelligence sources and methods
from unauthorized disclosure." The tap on Baldwin's phone,
meanwhile, remained active, but the White House had made no
explicit threats against Baldwin or the *Times,* so the newspaper's
publisher decided to forgo a major piece on the FBI's probe.

At the request of the White House, an FBI official met with
top Kennedy aide Kenneth O'Donnell to discuss what the bureau had found so far in its investigation (over 125 interviews) of

the Baldwin leak. The FBI had been providing O'Donnell with excerpts from the phone taps, which were passed on to the President. It was O'Donnell who finally had something of interest for the FBI: "The President and [I] have figured out that the finger of suspicion seems to point to one man." Kennedy wanted the suspect reinterviewed by the FBI, for this man had denied it previously. The President advised: Don't tell him that it was *my* idea to investigate him further.

The suspect? None other than Deputy Secretary of Defense Roswell L. Gilpatric. Number two at the Pentagon, he knew enough to provide the key information Baldwin had published, and he had met the reporter in D.C. on July 17. Exposing him must have been difficult for Kennedy. Unlike most others at high levels in this administration, Ros Gilpatric, born in 1906, had been handpicked by the President. A former corporate attorney, Gilpatric had served as under secretary of the Air Force in the early 1950s. Now the FBI would have to talk with Gilpatric again.

On the evening of August 22, President Kennedy met with CIA director McCone and General Maxwell Taylor at the White House. The meeting opened with an update on the new CIA office created by JFK's intelligence panel to control leaks. As per JFK's recommendation, high-level staffers who had access to sensitive material would now have to write up a memo on every contact with the press and submit it to their superiors. The new office shattered a provision of the CIA charter, under the National Security Act of 1947, that strictly forbade any intelligence activity within U.S. borders. Kennedy predicted that this would have "a very inhibiting effect" on officials talking with nosy reporters, knowing they were expected to pen a memo afterward. McCone observed that it could be set up so that it would look like the President was not "involved."

Then the discussion turned to another Cold War front: Cuba. Earlier that week spy planes had photographed a large number of Soviet merchant ships in the Atlantic. This brought to the fore a long-simmering debate within the administration. More than a year after the Bay of Pigs fiasco, the President continued to support various secret efforts—from propaganda to sabotage—to bring down Castro, but his top aides were divided on how committed the Soviets were to propping up the dictator. Some held that Cuba was an economic sinkhole for the Soviets; others that it was just a sideshow to the real flash point, Berlin. Still others pointed out that even if both those things were true, Cuba might be an important pawn for the Soviets—a vise to be squeezed until JFK considered a major trade: *Soviets out of Cuba, U.S. out of Berlin.*

John McCone had advocated some sort of "dynamic action" by the United States as the Soviets stepped up their military shipments. But what if the Soviets sent nuclear-tipped missiles? That would guarantee a far worse crisis. Robert Kennedy had argued that this was inevitable. Now McCone revealed that he had met with Rusk and McNamara the previous day to inform them that the "evolving situation" in Cuba appeared "more alarming." Photos showed the Soviets hauling to Cuba electronic equipment and large cases that might contain sections of fighter planes—or missile parts. Many of the ships "were unloaded very mysteriously at night in areas with all Cubans excluded from the area." Thousands of Soviets, both military and civilians, had disembarked from passenger ships at remote sites. An "atmosphere of secrecy surrounded the whole business."

McCone personally believed the Soviets were already constructing missile bases, but JFK cautioned, "We have to wait and see." The following day, to help "see," he ordered U-2 overflights, plus extensive contingency planning in case Soviet missiles did come to Cuba, "including blockade or invasion or other action." Kennedy also requested a study of American missiles near the Soviet border, in Turkey. Weren't they obsolete? Perhaps he was already pondering a possible offer if worse came to worst—trading them for removal of Soviet missiles in Cuba.

KENNEDY had told Rusk they just had to "ride through" the aftermath of the Fechter murder, but he remained curious about the episode. Acting for the vacationing McGeorge Bundy, George Ball sent a report to the President on August 24. It covered the stoning of the Soviets at length but offered only a few dozen words on the failure to aid poor Fechter: "The American Commandant was informed of the shooting approximately 23 minutes after it occurred. While he was in the process of determining what authorized action he could take, the body of the refugee was removed." End of story.

Meanwhile there was a new shooting victim at the Wall—a GDR border guard, Hans-Dieter Wesa, age nineteen. He had been assigned to a "ghost station" (where S-Bahn trains passed through but no longer stopped) near the border under a bridge. Wesa hated his transfer to the border police, as he opposed the idea of shooting anyone. He also had a sister in the West. On the night of August 23 he told his partner, who had become a friend, that he was heading out to switch on some lights down the track. He didn't come back.

His comrade ran after Wesa and spotted him a few feet past the Western side of a border-zone fence, which he had apparently scaled. As Wesa turned to flee, his partner fired six times and, after he fell to the ground, shot him in the face and body at close range, leaving a corpse a few feet inside the French sector. Mayor Brandt, fearing renewed rioting, visited the Wall that evening and expressed outrage while appealing for calm. Perhaps exhausted by the first-anniversary protests and then the Fechter episode, demonstrators stayed home.

AMONG those disturbed by Peter Fechter's death was one of the GDR guards who shot, and may have killed, him. Private Erich Schreiber, age twenty, penned a letter to his girlfriend, addressed to "My Dear Erika."

> *You write [asking] to know why I have been promoted. It is*
> *a more serious matter which does not happen to one every day.*
> *I have shot and killed a border violator who wanted to cross the*

*border from East to West. If that would upset you and you don't
want anything to do with a "murderer," please do not talk about
it with anyone.*

The letter was intercepted by the military censor and never
delivered.

Ten days after that murder, the victim was finally about to
be laid to rest. Fechter's family, however, was finding no peace.
Stasi agents had been harassing them for the past week. Peter's
mother, Margarete, sent a telegram to his beloved sister Liselotte in the West, informing her that "your brother will be buried Monday at 1200 hours." Liselotte told the press: "I won't be
able to attend my brother's burial. I fled East Berlin six years ago
and they'd arrest me if I went back."

On the eve of the funeral, the rabid East German radio
commentator Karl-Eduard von Schnitzler decried that when
criminal elements are "wounded directly on the border and not
retrieved immediately . . . then a huge fuss is made." He praised
"our brave young men in uniform" and sneered, "If you play
with fire you get burned."

Police tried to keep the funeral secret, ordering the family
not to post any sort of notice, but word had leaked. Three hundred citizens, who somehow learned about the funeral at a cemetery near Fechter's home in the Weissensee district, attended
the service. Some knew the victim from work, but most were
strangers. Fechter's family had wanted a religious service but
state officials forbade it, instead assigning a speaker from the
state's Municipal Funeral Commission. He told those assembled
that Fechter had made a "thoughtless" and "foolish decision."
Yes, everyone wishes to try different paths, he preached, but the
state wisely promotes and then monitors which paths they should
follow. East German citizens must respect that—"but Peter did
not." Peter's girlfriend and mother wept wordlessly. The latter
had spent her entire savings for a mahogany coffin and wreaths
for the funeral and had to borrow money for a tombstone, in-
scribed with *Unforgotten by All*. After the crowd left the cem-

etery, Stasi agents removed all of the flowers the mourners had left at the grave.

Five Western journalists (three Americans, two British) would be detained after trying to cover the service, their cars pulled over as they left the scene. Among those taken to police headquarters in Alexander Platz were reporters or photographers from *Life* magazine, the London *Daily Mail*, and the BBC, as well as the Associated Press's Berlin bureau chief. Police confiscated the film of two photographers. An officer lectured one, "The Western press by its activities is disturbing the peaceful citizens of West Berlin but it may not disturb the peaceful citizens of East Berlin." The U.S. Mission described it this way in a cable to Washington: "The newsmen were accused of committing an unfriendly act against the GDR regime by covering the funeral." The foreign journalists were released after three hours, but a West German photographer was kept in jail, charged with the serious crime of "transmission of information."

In America, *Life* published a seven-page article on the Fechter murder, titled "The Boy Who Died on the Wall." Its Luce-owned sibling, *Time*, went one better, with a striking illustrated cover: an arm reaching under barbed wire atop a concrete barrier on which hung a wreath of flowers for the latest victim—titled simply "The Wall." The accompanying cover story, "Wall of Shame," depicted West Berliners as normally cocksure types who "thumb their noses at trouble," even though hardly a night passes "without the rattle of gunfire and the sounds of death from the other side." But after the latest very public shooting they had gone on "an emotional bender. . . . Suddenly, all the pent-up frustrations exploded" in "an orgy of riots," aimed at Germans and Americans alike. For the first time ever, cries of "Americans go home" were heard. "The mob's voice echoed in every major capital in the world, forcing Russia and the West into another of those nightmarish Berlin confrontations," the writer continued. "It emphasized once again that so long as the Wall is allowed to stand, a perpetual threat to world peace exists in the heart of Europe."

Few Berliners believed the Wall would fall in their lifetime, *Time* reported, and reunification remained only a "remote prospect." A West German official, while friendly to the United States, complained that the "threat of nuclear war has paralyzed the West." But the article concluded that at some point—within a decade or a generation—the Wall "must come down," and "if it is not brought down by reason, will someday provoke men to demolish it by force."

PRESIDENT Kennedy seemed determined to double down on the approach that had recently shaken America's standing in Berlin: focusing completely on the West, while considering East Berlin, as the U.S. soldier said on the day of the Fechter murder, "not our problem." Mike Mansfield, the majority leader of the U.S. Senate, had just written JFK a memo suggesting more proactive American action. Kennedy replied, "I think the truth is that East Berlin is not an issue of war and peace for us, and therefore we should not adopt any of the more drastic alternatives proposed in your memorandum. The time for a fight over an effective Western role in East Berlin passed, if it ever existed, many years ago. . . . The crucial question remains that of West Berlin."

On August 29, besides announcing his nomination of Arthur Goldberg to the U.S. Supreme Court, Kennedy met with Rusk, Bundy, and others to talk about Berlin for the first time in three weeks. They agreed that the United States should continue to restrict Soviet entry to West Berlin, even if this cost American access to the East. Bundy said flatly, "We have nothing left in East Berlin that's of any interest to us."

Later in the discussion Kennedy affirmed, "Well, we don't care about East Berlin, we know. So the [access] rights—we just aren't interested in."

And what was the mood in the sector the administration did care about, West Berlin? What a "strange and frightening week," Sydney Gruson, the Berlin correspondent for the *New York Times,* declared in an insightful commentary. It had left the nerves of the West Berliners raw and "sharpened their anxieties and sorrows." With the Fechter killing, "[s]omething snapped in the West Berliners. . . . It was as though a piece of string had been stretched too taut for too long," leading to almost unbelievable scenes of West Berliners taunting their own police and telling Americans to go home. Gruson, a close friend of Daniel Schorr (and husband of foreign correspondent Flora Lewis), felt that emotions had overtaken reasoning, yet with trauma had also come a certain clarity. He observed:

> More than any single event since the wall was built, Peter Fechter's lonely and brutal death has made West Berliners feel a sense of helplessness in the face of the creeping encroachment being worked so subtly by the Communists. The city feels alone and cut off, not so much by the 100 miles or so of Communist-controlled territory between it and West Germany but by the inaction forced on people when they want to do something, almost anything, forceful to counter the Communist measures.

Even as older folks warned that things could get much worse if the Communists were punished, younger people were "willing to act," according to Gruson. After Fechter's death, and the riots, "there is less hope than ever that the West Berliners will live with the wall. So long as the wall is there, some among those trapped behind it will probably try to escape. So long as they do, there will probably be killings by the East German guards and so long as there are killings the seeds of explosion will remain in West Berlin."

As the article appeared, three of those younger West Berliners who had been "willing to act" were getting ready to go on trial in East Berlin, knowing they faced certain convictions and lengthy prison sentences.

12

Coming Up Short

The show trial for three Girrmann couriers captured on August 7 was slated for two days of testimony, after which a panel of judges at the Supreme Court building in East Berlin would announce the prisoners' preordained fate. The accused— Wolf-Dieter Sternheimer, Hartmut Stachowitz, and Dieter Gengelbach—had endured endless interrogations and admitted to at least some aspect of their alleged crimes. Now their state-appointed lawyers were entrusted to make sure the prisoners memorized their lines. They were to confess that they had been affiliated with the Girrmann *terrorgruppen* and had seduced East Germans to leave their nation illegally. They were to confirm, in addition, that Girrmann was funded and directed by the West German government, the West Berlin Senate, and German and American intelligence agencies aided and abetted by someone named Mertens. These overlords had promoted an "armed clash" with the possibility of killing East German guards in cold blood to "bring untold suffering to the people of Europe and primarily upon the German people," according to the prosecutors.

Citizens were allowed in the courtroom, but they were limited to Communist supporters from wide-ranging fields who could carry the results and message of the trial to their co-workers. For much the same reason, journalists from East Germany and other Communist bloc countries were also present in high numbers. The president of the Supreme Court of the GDR, Dr. Heinrich Toeplitz, presided. Among the exhibits were a pistol and bullets allegedly linked to the tunnel operation and a photo of the rectangular hole cut out of the Sendlers' living room floor.

The three couriers in the dock were easy to distinguish. Sternheimer was blond and thin-faced, Gengelbach dark and paunchy, Stachowitz with a receding hairline that made him look older than his twenty-six years. Witnesses included a truck driver approached about joining the plot. An East Berlin woman claimed she had been repeatedly visited by couriers urging her to escape with her two sons. One might speculate, although not in that courtroom, that she must have once expressed interest in the idea. Now she claimed she spat on the invitation. When one of the couriers had directly asked her son, who was fifteen, if he wanted to flee, she was "so shocked" that she "threw out the intruder." Manfred Meier, who would be tried separately, repeated his TV confession while again refusing to state that his fellow *fluchthelfer* had planned to shoot heroic GDR guards at Kiefholz Strasse.

Edith Sendler told the court that on escape day she was startled when she heard the commotion under her floorboards. Her husband was in the house at the time, she falsely testified, and was so upset he had an asthma fit. (Stasi files, however, continued to describe the Sendlers as "scheduled for smuggling," their expression for *wanting to flee*.) To Sternheimer, sitting in the courtroom, Frau Sendler seemed deeply distressed, and he guessed that she, like others, had been interrogated at length. Had she agreed to shade her testimony in exchange for leniency for herself and her husband? Perhaps the Stasi felt it was more important to portray the tunnel as an unwelcome invasion, rather than an invited one, in order to make the defendants even more evil—"kidnappers" and "abductors," in GDR parlance.

On day two of the trial, Stachowitz recalled meetings at the

House of the Future. "A woman with the group told me that there is a close connection to U.S. intelligence," Stachowitz said, his lines memorized well. He disclosed that West Berlin police promised to lay down "covering fire" if the tunnelers or couriers got in a fix. Gengelbach testified that he had met with the notorious LfV operative Mertens to ask for weapons and money, although it was unclear if he had received either. The name of celebrity *fluchthelfer* Harry Seidel was linked to Mertens as well. Gengelbach claimed that *Bild Zeitung* had given him 1000 DM, and an unnamed American TV network (no doubt CBS) 800 DM, to chronicle his part in the escape. The prosecutor declared that Gengelbach went to the East several times, beginning in July 1962, "to establish contact through other persons to the owner of the carpentry shop where the tunnel was supposed to end. Those efforts were not successful."

After three days the expected verdicts from the judges came down: all were guilty on all charges. *Neues Deutschland* reported the judges' claim that this "righteous judgment" was "given in the name of the people" and "should be a warning to all those who think they can touch the security and sovereignty of the GDR." Gengelbach got twelve years in prison and Sternheimer eight. Stachowitz, who still did not know where his wife and son were, expected worse than his six-year sentence. When he got out of prison, he now knew, his son would not yet have reached his tenth birthday. That was something to hold on for.

WITH the conclusion of the show trial, Edith and Friedrich Sendler were finally released from custody. Edith's testimony (truthful or not) was hardly necessary to convict the three Kiefholz tunnel couriers. Perhaps Herr Sendler had offered to secure for certain officials some of the same black market luxuries he had a knack for obtaining—and/or to act as a Stasi informer, as his position of tradesman afforded him wide contacts across GDR society.

On the evening of their release, the Sendlers were accompa-

nied to their cottage by a Stasi lieutenant, who returned some
of the items plundered from their home by soldiers and MfS
operatives. The Sendlers must have been shocked to find that
so much, including pricey jewelry, was still missing, but they
began to settle in. Then: a startling turn of events. Shortly after
six o'clock, shots were fired from the West over the fencing at
the house. The Stasi operative counted eleven shots in this initial burst. Every time he pulled back a curtain to take a look,
another shot rang out. In the end no damage was done, so it was
hard to say if these were warning shots or in deadly earnest,
but they indicated to the Stasi that the house had been under
observation awaiting this very moment. For how long? Almost
a month had passed since the botched escape. Did someone in
the West have inside information on exactly when the Sendlers
would return home?

The Stasi report on the incident observed: "During this time
period a male person moved behind the border installations [in
the West]. No details could be identified because he stayed behind a board fence and darkness fell." The couple demanded that
they be taken from the house and left shortly after seven, evacuated by a Stasi staffer to the nearby home of Friedrich Sendler's
mother. "When they left the house," the report related, "no shots
were fired." At eight the following morning an officer accompanied the Sendlers back to their home, and left them there to fend
for themselves.

THE future looked bleak to Gerda Stachowitz, who had heard
nothing of her husband's trial, nor where her son, Jörg, was living. Still, she tried to keep up her spirits and those of friends
and family members. As she got ready for her own trial, she
attempted to smuggle short messages (known in Stasi-speak as
"stiffs") out of the prison, which were, of course, intercepted by
the authorities.

The first one revealed that she was still in prison: "Do everything for us, go district attorney. Watch out! Walls have ears.

Gerda." The second: "Where is Jörg? Situation very serious. Stay healthy. You are my hope. Clean up everything. Gerda." Another: "Please call number XXXXX—ask for Mrs. [redacted]. Read our letter 2x. Many, many thanks. Shall write Ma or pick up letter!" The Stasi checked the phone number and found it was based in Berlin-Friedrichshagen. Gerda finally managed to get a longer note out of the prison, but it, too, was apprehended by the Stasi. And it was a heartbreaker:

Dear loved ones!

Today is [redacted] birthday. Don't despair. It can take a long time until everything is good again. We have gotten into deep trouble. Don't think about it but pull through so that we most certainly can meet again healthy.

Don't get mad at us. Especially my beloved mommy.

The [prison] sentence need not be the same as the actual length of the punishment. You won't be able to do much anymore at this point. Before the trial, go to the attorney general who leads the trial—good reputation/characters might be useful. Maybe it can be achieved later that we can work in our old professions.

I hope so much that one day I can be happy again with Hambi [her husband]. We didn't work for anybody, only had connections, so that I'd be able to get away. Get a good defense lawyer who is familiar with the problems.

We want to build up each other's courage and spirit. The important thing is that we are all still around. My thoughts are always with you [all] and my beloved husband.

Love, forever your Gerda

Those interrogating Manfred Meier, meanwhile, were growing infuriated with him. Their sessions now stretched for four hours or more, in advance of Meier's own trial. Questioning would typically begin with the Stasi officer opening a window to let in fresh air and the pleasant sounds of birds chirping. "Meier," he might say, pleasantly, "you are a cool fella, such an intelligent guy, don't make this so difficult—it's pointless. Just tell us what you know." All the MfS wanted was for Meier to admit that his heavily armed fellow escape helpers had planned a bloody attack on the East. As Meier denied this, over and over, the window would be slammed shut. The Stasi man would bring his face up to Meier's and shout, spittle flying, while banging his fist on the table, "Do you really think we are all stupid here? Damn it, tell us the truth about the weapons!"

After more threats and bullying, the window would be opened again, and this cycle would be repeated, for hours.

AT 2:55 on the afternoon of September 4, an entry was made in the U.S. Army's Berlin Brigade intelligence log: "At Berg Strasse, corner of Bernauer Strasse, 5 m within cemetery, E Germ civilian shot, appeared fatally. Removed on a stretcher by VoPos." The victim was Ernst Mundt, a forty-year-old former construction worker on a disability pension. When the Wall separated him from his relatives in the West, he had grumbled about it from the start. Finally he decided to do something about it.

That afternoon Mundt rode his bicycle from his apartment in Prenzlauer Berg to the highly restricted Sophien Cemetery at the Wall just off Bernauer Strasse. Wearing a dark cap on his head, he climbed atop the cemetery wall perpendicular to the barrier at the border. It was covered with sharp glass to discourage just such a move. Then he ran toward the Wall, shrugging off pleas by onlookers. "I *won't* get down," he shouted. As he was about to reach the Wall, where a good leap might carry him over, two East German transport police officers about a hundred yards away took notice. One of them fired a warning shot, and then took deadly aim. A bullet ripped through Mundt's head and he toppled over, just feet from freedom. His cap flew over the Wall. West Berliners found it, with a bullet hole through it. Mundt quickly became known to them as "the Man with the Cap."

The following day the policeman who shot Mundt was awarded bonus pay and a Medal for Exemplary Service at the Border. He had "handled his weapon superbly and put it to use masterfully." The troop leader in the area was commended for removing the slain criminal before the West Berlin police, press, and camera crews could arrive. This was in the wake of new orders, following the Fechter incident, that bodies be hustled away immediately to prevent protests and news coverage in the West. Nevertheless, hundreds of angry West Berliners vented their fury across the border that evening, erecting a cross decorated with flowers near where Mundt's cap had landed.

A couple of blocks down Bernauer Strasse, under the swizzle stick factory, the digging went on as usual. Spirits were high with a finish line in sight. What the tunnelers dreaded most, perhaps because they had already experienced it, was another water leak. So when they noticed a puddle on the floor of the tunnel near where they were digging, and moisture dripping from the ceiling, they took notice. It might be nothing. Perhaps rainwater had collected in a low spot overhead or they were under a building with poor plumbing and the nuisance would depart as quickly as it arrived, with no explanation.

Then the drip turned into a steady leak.

On the plus side, there was hardly enough water on the floor to warrant an immediate work stoppage. On the other hand, unlike with their previous leak, they could not ask sympathizers in the West Berlin utilities department for help, as they were now well past the Wall. There was no one in the East to fix a broken sewer pipe, and they were so far from the West that any pumping operation would be difficult. They could dig on and pray the leak stopped or at least didn't get worse, but they didn't expect to reach their destination under the Bulgarian's tenement on Rheinsberger Strasse for three weeks or so. That seemed far too long to court disaster.

So the tunnel originators, plus their top lieutenants, decided to explore the possibility of coming up for air early, in another East Berlin basement . . . god knows where.

THE question of whether Soviet missiles were heading for Cuba had rapidly become critical to President Kennedy—and, in his mind, critically linked to Berlin. But in the early days of September another crisis demanded his attention, causing him to cut short a Newport vacation. An American spy plane had briefly entered Soviet airspace over the tip of Sakhalin Island, breaking the ban on overflights the White House enacted after the U-2 shoot down of 1960, which led to the capture of pilot Francis Gary Powers. This time the Soviets didn't fire at the high-flying jet, but they spotted it on radar and protested loudly. Another high-flying U-2 was still taking photos over Cuba, so the White House didn't want a new spy plane crisis to erupt.

As the late morning meeting on the straying U-2 commenced, Dean Rusk said, "It's very clear indeed that the Soviets have got us right on the hip on this one." The plane had simply drifted off course for about nine minutes at night. Still, the United States didn't want to get Soviet leader Khrushchev excited—he might do something rash regarding Berlin. So Rusk read a draft statement falsely calling the U-2 a "weather reconnaissance and air-sampling aircraft" that had "unintentionally" been victimized by Mother Nature.

"It undoubtedly did *some* air sampling, didn't it?" Rusk asked, hopefully. "Don't all our flights do some of this?" Others around the table indicated: no way.

"Well, I don't know . . ." Kennedy replied, "we don't owe him [Khrushchev] the whole truth." The President argued that mentioning "night time" would indicate no photography. "That seems to me—that gets away from the U-2 idea."

"But it *is* a U-2," Bundy reminded him. Kennedy nevertheless decided that calling it a "weather reconnaissance aircraft" without the "air-sampling" detail might be enough. (This was what the United States had claimed for weeks after the enemy shot down Powers's spy plane—until the Soviets produced the pilot.)

Some in the room exited, leaving JFK, Bundy, Robert Kennedy, McNamara, and Rusk to take up the Cuba question. The President had to respond to rumors in the press and claims by Republicans that nuclear missiles had *already* arrived on the island. He urged caution, but two of his cabinet officers wanted to discuss what to do if and when the CIA confirmed that surface-to-surface missiles were indeed in Cuba.

"I think we'd have to act," Rusk declared. "For example, I would suppose that if you're going to take on a bloodbath in Cuba, you'd precede it by a systematic blockade to weaken Cuba before you actually go to put anybody ashore."

"See, I wonder why . . . if we'd do it then," McNamara replied, "why wouldn't we do it *today*? This is one of the actions that we can consider today, as a matter of fact. There's no question the Soviets are shipping arms to Cuba, that's clear. They said so. Now we can——"

The coolest person in the room interrupted. Fortunately that person was the commander in chief. "The reason we don't is that—is because we figure that they may [then] try to blockade Berlin," Kennedy observed. That would immediately cause a horrific crisis for the United States while a blockade wouldn't do much harm to Cuba "for quite a while." That settled it, for the moment.

Later that day, Kennedy briefed congressional leaders. He outlined what he knew about the Soviet buildup but added that "even though I know a lot of people want to invade Cuba, I would be opposed to it today." Again he invoked Berlin. After any U.S. move on Cuba, "Berlin obviously would be blockaded also." And what timing: "Listen, I think Berlin is coming to some kind of a climax this fall, one way or another. . . . We have to weigh our dangers. I would say the biggest danger right now is for Berlin."

The meeting ended with Kennedy requesting standby authority to call up another 150,000 reservists. Press Secretary Pierre Salinger put out JFK's official statement on Cuba, warning that the "gravest issues" would arise if the Soviets introduced offensive missiles there. What the White House did not know was that the first Soviet shipment of medium-range R-12 missiles, capable of carrying a thermonuclear device, was about to reach Cuba by sea. Another was expected in mid-September. Anticipating new warnings from the White House, Soviet forces would go on their highest alert ever.

Premier Khrushchev, confident he would get away with his Cuba move, grew increasingly aggressive on Berlin once again. He abruptly summoned the closest U.S. official at hand—a very surprised Secretary of the Interior Stewart Udall—to the Black Sea resort where he was vacationing to inform him that after giving Kennedy a pass for the upcoming midterm elections he would offer JFK a choice: "go to war or sign a peace treaty" ending the American occupation of West Berlin. "It's been a long time since you could spank us like a little boy—now we can swat your ass," Khrushchev boasted through Udall. Surely the missiles coming to Cuba were on his mind. "We're equally strong," he said. Pointing out how deeply Berlin sat in East Germany, he added, "We have the advantage. If you want to do anything, you have to start a war."

The Soviet leader also met with a far different visitor—America's most famous poet, Robert Frost. He told Frost that Americans were "too liberal to fight" now, recalling (in his usual earthy manner) Tolstoy's comment to Gorky on aging and sex: "The desire is the same—it's the performance that's different." In a public statement, the Soviets declared, unequivocally, that they had not and would not send nuclear missiles to Cuba. President Kennedy continued to tell his aides that the coming crisis he feared most would be over Berlin.

AS water continued to gather in puddles on the tunnel floor, organizers of the Bernauer tunnel knew they had to make a quick decision on cutting their path short, but first had to find out exactly where they were in East Berlin, no easy task. They hadn't done any surveying for a while, so Joachim Rudolph summoned his engineering buddies Uli Pfeifer and Joachim Neumann to take new measurements. With their theodolite mounted on a small tripod, they measured angles and charted the course on their map of the target neighborhood in the East. Miraculously, after all this time, detours around boulders and other zigzags, the engineers found they were still almost perfectly on track to that cellar on Rheinsberger, although it was still almost one hundred feet away.

Plotting how far they had gone along that route, the tunnelers discovered that—if their measurements were correct—they were now nearly below a tenement on Schönholzer Strasse, one block closer to the Wall than their original target. The building was marked as #7 on their map, which also indicated that it had a basement. Perfect, except that they had never seen the tenement, let alone its cellar, did not know anyone in the building or on that block, and had keys neither to the building nor to the door to the basement.

Joachim Neumann had one other very personal concern. He had planned to usher his girlfriend, Christa, through this tunnel. They had been corresponding very carefully for months, knowing the Stasi was probably reading the letters, or sending coded messages through relatives. (Neumann would write, for example, "I got the message from your aunt, I am very glad and I will see what I can do.") Christa was set for the late September escape date, but now her boyfriend learned that she was to be on holiday for most of the month, and so was certain to miss an early breakthrough. "We have to take our chances and keep digging!" Neumann argued briefly, before painfully admitting that they had no choice but to wrap it up as soon as possible, Christa or no Christa.

A courier who was able to cross the border inspected Schönholzer Strasse and peered into the lobby of #7. He returned with a new concern: The street was subject to strict police controls. Buildings on the first block beyond the border had been torn down or evacuated to prevent escapes, so the near side of Schönholzer now held the housing closest to the death strip. Residents and guests on that side had to show their papers to guards at checkpoints, with gates at either end. Barbed wire ran the length of the street right down the middle. On the 7 Schönholzer side, slightly farther away from the Wall, residents and guests could pass freely—but with guards and police on patrol just across the street. Refugees unfamiliar with the block, some with baby carriages, highly stressed and searching for street signs and building numbers, were almost certain to attract VoPo attention.

Coming up short on this particular block was especially risky, maybe crazy—but insanity born of desperation and urgency. And with the water leak intensifying, there was no time to waste.

WITH security a more central issue than ever, the handful of tunnelers who knew about the revised schedule kept this rather critical news from their colleagues. They needed their comrades to dig those final yards to the new target; then, very close to the breakthrough, they would learn the truth. This would leave a small window for them to contact those on their escape list, but security trumped all at this point.

They did alert NBC, however. Piers Anderton quickly sent a coded message to Reuven Frank in New York that he might have to fly on short notice to Berlin if he wanted to be there for the big event. In a crucial step, Anderton's boss in Germany, Gary Stindt, decided they needed another camera position for the final days, so he rented an apartment on a high floor at Bernauer and Wolgaster across the street from the swizzle stick factory. From there a camera manned by NBC veteran Harry Thoess could overlook checkpoints and guards, rows of boarded-up buildings, and East Berliners walking and talking on the streets. And most important, thanks to the GDR's destruction of buildings close to the border, there was a clear sight line across the Wall to that newly crucial block of Schönholzer Strasse. And 7 Schönholzer, its front door flanked by small basement windows, was standing right there in full view.

Back at the swizzle stick factory, the tunnel's clay floor was slowly turning to mud beneath the wooden planks. The engineers were fairly confident they were just a few feet from their new target, but uncertainty abounded: What was the cellar floor made of and how thick was it? Would the breakthrough take minutes or hours? What was the layout in the basement? Where did the steps come out in the lobby, and was that door kept locked? Was the cellar lightly or heavily used by tenants—was anyone likely to confront them as they chipped through stone or waited for refugees to arrive?

For now these would remain nagging mysteries. What they had to concentrate on—a challenge they had thought was weeks off—was lining up couriers at short notice to cross the border and alert refugees on or before escape day, now set for September 14. Hasso Herschel and Uli Pfeifer took charge of this. A quick check with young West Berliners who had previously expressed interest in serving as couriers produced discouraging results, with several now out of the picture (they could not be found), sick (real or feigned), or too busy (ditto). Fortunately, a new candidate for heroics had arrived on the scene.

This was Ellen Schau, fiancée of Mimmo Sesta. By chance, she had arrived in Berlin from Düsseldorf, where she worked as a secretary in a patent attorney's office, on September 10 to spend the week of her twenty-second birthday, which would fall on . . . September 14. Mimmo and Gigi picked her up at the airport and unveiled the courier idea over dinner.

Ellen didn't look like your typical escape agent. Wearing tasteful jewelry, her reddish hair in a French twist, she looked elegant, even chic. Until that moment she wasn't even aware of what her boyfriend had been doing the past six months, let alone that the climax was approaching. True, Mimmo had been acting oddly since March, incredibly busy, often hard to locate. When they did talk on the phone he didn't want to discuss his studies. He postponed trips to Düsseldorf. Now he informed her she was perfect for the role of courier. She had a West German passport. Stasi agents and police officers would not likely be suspicious of an attractive young woman, and if challenged she could produce papers showing she was just another out-of-towner. On the other hand, she had never been to East Berlin, didn't know the signposts or streets at all, and had a phobia about riding on the U-Bahn and S-Bahn trains.

Despite feeling scared and unprepared for the mission, she accepted it.

That night Ellen stayed on Ansbacher Strasse with the girlfriend of one of the diggers. The next day at a Technical University dorm she was introduced to some of the others involved in the operation. There she learned about the route she would follow, and the names and addresses of bars and cafés where she would transmit coded signals to refugees gathering at the appointed hour. The organizers would give her a wad of cash in case she felt it was too risky to try to return to the West right away.

Meanwhile, Stindt and Anderton had agreed to let the tunnelers use the apartment NBC had rented overlooking Schönholzer as their off-site headquarters on escape day. This contravened Reuven Frank's order that his men were not to assist the tunnelers in any way (beyond funding them, of course). Someone on that perch would communicate with those in the basement, alerting them to any danger visible through binoculars—also provided by NBC—across the Wall. Whoever was in that apartment would hang a sheet out the window: white if the coast was clear, red if Ellen Schau and the other couriers should warn the refugees to return home.

WEDNESDAY, *September 12 (two days until breakthrough)*. Ellen and Mimmo shared breakfast at Cafe Bristol, again tracing her routes to the cafés and pubs in East Berlin where she was to signal refugees. Her retreat in the East, where she could pass time between those visits, would be historic Zion Church, where the Lutheran pastor Dietrich Bonhoeffer had preached in the early 1930s before he began organizing resistance to the Nazis (in 1945, he was executed for this). Since she didn't want to get caught by the Stasi with a map on escape day, she returned to her room on Ansbacher and began memorizing her routes and signals.

Mimmo Sesta, meanwhile, crossed the border to notify Peter Schmidt in suburban Wilhelmshagen about the surprisingly close escape date. From the NBC apartment in the West, high above the checkpoint, Harry Thoess filmed Mimmo, in his favorite trilby hat, making his way through the checkpoint to the East. Shrugging off their near-arrest a month earlier, the Schmidts were happy to hear they would get another chance to flee.

Anita Moeller had not spoken with her brother, Hasso Herschel, since the Kiefholz bust. Knowing her brother, she was certain he was working on another escape, and she was desperate to exit, with or without her husband, who now wasn't so sure he wanted to leave the East. They were separated again. She and their daughter, Astrid, were still staying with her parents in Dresden, while Hans-Georg worked as a construction engineer in Senftenberg, fifty minutes north by car. He told her that whenever a stranger came into his office he worried, *This is it, they know about Kiefholz Strasse, and I will be arrested like the others.*

Given a choice of patching things up with her husband or leaving him behind, Anita knew she would choose the latter, for her own sake and that of her child. So when she received, for a second time, a coded telegram urging her to come to Berlin immediately, she packed up a couple of things for the baby and a most cherished luxury: her black Dior wedding dress. Bidding farewell to her sister, Anita said, "I don't know how it will work out." Her sister offered some practical advice: if the VoPos caught her on the way to the site, "just tell them you'd *never think* of going into a tunnel—because of your claustrophobia." Anita didn't explain to her parents why she was leaving as she bid good-bye at their door.

Learning of this second chance, her husband decided to drive with her to Berlin the next day. They would stay overnight with a friend. He could decide whether to stay or leave on the morning of the escape.

THURSDAY, *September 13 (one day until breakthrough).* After breakfast, Mimmo and Ellen took a tram to Bernauer Strasse. Mimmo pointed to the window high in the sky where he or a comrade (or maybe even Harry Thoess) would hang the white or red sheet out the window of the NBC flat. Then they climbed to that apartment for a look across the Wall—and at the battered tenement at 7 Schönholzer. Ellen returned to her room for more studying.

Hasso Herschel, meanwhile, led a team of diggers in preparing for the breakthrough. Shoveling and using a pick, they had the difficult task of digging upward for several feet at a 45-degree angle until they hit something solid. Which they did, with a loud *clank* and considerable relief. Now they were set for the final push the following day. One concern: they were not certain the building they were actually under was 7 Schönholzer. Another: for all they knew, the Stasi had been tipped off (if not by Claus Stürmer, by someone else) and armed agents were already waiting for them on the other side of that cellar floor.

While this was happening, the tunnel organizers met to go over assignments for the following day. Their fellow diggers, who still knew nothing about the breakthrough, would be told to come to the factory in late afternoon for a "special meeting." There they would be assigned posts inside the tunnel, spaced every few yards, to help usher escapees along the way. Some would be stationed in "cut-outs" where any refugee (some would be infants) on the verge of panic or having second thoughts could be pulled off the center planks for a few moments, allowing others to crawl past. A couple of the project leaders would remain at the basement in the West to greet the escapees and get them loaded on the van to their new life.

But who would conduct the absurdly risky breakthrough itself, and then wait in the 7 Schönholzer cellar in the East, perhaps for hours, for refugees to arrive? At a meeting of the top echelon, Joachim Rudolph felt deep ambivalence, the memory of his near-death experience in the Sendlers' cottage still fresh in his mind. He was relieved, but hardly surprised, when Herschel again volunteered to lead the team into the East. This gave Hasso the authority to pick his mates. And sure enough, he wanted Rudolph and old friend Uli Pfeifer at his side again. Hasso knew he could count on them; they had refused to panic in the Kiefholz operation, passing a very strenuous real-life test.

Rudolph, despite his fears, accepted, but Pfeifer was out of the question. His mother, after hearing about Kiefholz, had made Hasso promise not to put her boy in that kind of danger again! So Hasso selected another tunnel veteran, Joachim Neumann. As one of the originators, Gigi Spina exercised his right to join them. (Mimmo would supervise at the Western end; Wolf Schroedter's recent medical calamity would limit him to a supporting role.)

One more person needed to be contacted about the operation. Although some now regretted making that promise back in the spring, they had little choice but to invite on this adventure, however briefly, Christian Bahner, the son of the man who had donated so much lumber to the project. His good intentions could not be questioned, but he had alarmed them with his naïveté and foolhardy behavior on his last visit, when he had waved around a shotgun and displayed sticks of dynamite, as if ready to lead a one-man insurrection in the East. Gigi Spina considered him a "fanatic" and told him, "Man, you are crazy!" But that lumber had saved them thousands of DMs, and a promise was a promise. They would have to invite him to enter the basement in the East, for a few moments before shooing him back to the West.

When Hasso's sister Anita arrived at her friend's apartment, husband and infant in tow, she vowed not to repeat the too-festive night before the Kiefholz tunnel escape. Since she was barely even speaking to her husband these days, Anita was hardly in the mood anyway. That evening, Gigi visited Mimmo and Ellen over on Ansbacher Strasse. Gigi revealed that Claus Stürmer would not be allowed to leave the swizzle stick factory that night and would be kept under guard going forward. In about twenty-four hours, if they pulled off their miracle, he would be allowed to send a courier to his wife in the East.

IN New York Reuven Frank had received the long-awaited call from Piers Anderton: it was time to get to Berlin posthaste. He knew it could only mean one thing, but he couldn't guess how long he might have to stay. Delays, given what had happened so far, might push the climax back days or weeks. Well, at least he would be nearby to offer advice—and try to keep his own staffers out of trouble, if possible. He ordered one of his top film editors, Gerald Polikoff, to cancel what he was doing and join him on a flight across the ocean. This made Polikoff only the fifth person at NBC headquarters to know about the project.

Landing at Tegel Airport in West Berlin after a twelve-hour journey, they were retrieved by Anderton, Stindt, and cameraman Harry Thoess. Frank was informed, "They go through tomorrow night—the tunnel's finished." Then they drove him past the swizzle stick factory on Bernauer Strasse for his first glimpse of the tunnel's home. Frank realized that he had stood just a block away with David Brinkley the morning the Wall went up.

That afternoon, in the NBC office on the fashionable Kurfürstendamm, the three German-based staffers started showing Frank some of the twenty hours of footage. It amounted to about 12,000 feet, developed in high secrecy by a film processor in Berlin. Frank was impressed by the footage Anderton had assembled from his early visits to other escape sites, from sewers to early tunnels. But he was stunned by his first glimpses of the Bernauer tunnel, going back to its opening weeks. Anderton identified the key characters: "the Italians," someone named "Hasso," and a "Wolf." Then there was "home movie" footage shot by Sesta on visits to the Schmidts.

Frank was thrilled. It went far beyond what he had expected for a documentary exploring a full year of Berlin escapes. This was infinitely more extraordinary. Most TV coverage was mere "news," minutes or days after the fact, and sometimes you had to be lucky to get even that. This was history in the making, cinema verité, danger at every turn, day after day, happening right in front of the camera's eye—one might call it something new for TV, a reality show filmed at the front line of the Cold War. Frank knew journalists who had spent a lifetime without achieving this, and now NBC, he felt, had practically stumbled into it.

13
Schönholzer Strasse

SEPTEMBER 14, 1962

The morning dawned warm and sunny, promising a gorgeous late summer day in Berlin. If they wanted to take this as a positive sign, the tunnelers could, and some did. Couriers and refugees, at times pushing baby buggies, would not have to dodge raindrops and carry umbrellas on the way to the tunnel.

Joachim Rudolph brought a few new tools to work to help punch a hole into the basement of (what he hoped was) 7 Schönholzer. The breakthrough was set for late afternoon. Rank-and-file tunnelers, invited to what they thought was just another meeting, would not arrive until after 5:00 p.m. Mimmo Sesta and Hasso Herschel huddled for the final time with the couriers, principally Ellen Schau and the Swedish girlfriend of one of the diggers. Ellen, who was slated to meet groups of refugees at three pubs, decided to cover her striking red hair with a scarf to draw less attention. Since she had flown to Berlin that week to celebrate her birthday and dine at some nice restaurants, she had not packed any informal clothes, so she borrowed a light overcoat from friends. Mimmo handed her the promised cache of DMs in case she got stuck across the Wall.

Over in the East, the Schmidts awaited confirmation that this was, indeed, escape day. Peter's mother, who worked in a menial position at a Soviet commander's office, had arranged for the day off. Anita Moeller still did not know whether her husband would join her in fleeing to the West. Recalling the last escape, Anita decided not to give her daughter, Astrid, part of a sleeping pill, taking her chances with a squalling toddler. Hans-Georg walked her to the tram, trying all the way to persuade her to stay in East Berlin. When they arrived at the tram stop and she stood firm, he said, "Okay, I will leave with you."

Back at the NBC office, Reuven Frank continued watching the hours of footage shot in and around the tunnel, growing more excited, and nervous, by the minute about that evening's escape. Around noon he called his boss, Bill McAndrew, back in New York and told him, "I think we'll need ninety minutes for this program, but I'll know more tomorrow." His cameraman Peter Dehmel was out in the streets, filming Ellen Schau as she made her way, wearing dark glasses, her hair still in a French twist but wrapped in a scarf, to the S-Bahn's Zoo station. Ellen's phobia about riding the elevated train was held in check only because she had so much else on her mind. Dehmel shot her in a nearly empty train car until it made its last stop at the border. Then he got off, and tracked the train with his camera for a few minutes as it proceeded into East Berlin. To enshrine the time, he filmed the clock at the station. It was almost two o'clock.

Another courier set off for the Schmidts' home on the outskirts of the city to give them their final instructions. A month earlier, not knowing that the escape at Kiefholz Strasse was in the works, Eveline Schmidt had been running errands when the courier arrived. This time, thanks to Mimmo's visit two days earlier, she and her husband had remained at home all day.

Once the courier left, Peter Schmidt rushed to notify his mother. Returning home, he donned several layers of underwear, figuring that if he got arrested at least he'd be prepared for weeks or months in a cold, dirty jail cell. Eveline put on a new dress and packed a shoulder bag with two extra diapers for daughter Annett. Then they walked away from their little home, hopefully forever, to catch the S-Bahn to Alexander Platz. The Schmidts were on edge already, but they might have been even more nervous if they had known that in recent days Peter's mother, anticipating escape day, had sold some of her furniture, usually considered a tip-off to authorities that the homeowner was getting ready for a major move. And *she* worked at a Soviet headquarters.

By then, Ellen Schau had arrived at the pub just blocks from 7 Schönholzer where she hoped to signal the first refugees. Realizing she was quite early, she walked to the Zion Church nearby, sat, waited, and prayed. She had memorized most of her routes, but carried a few notes in code as backup. Now she destroyed them. With more time to kill, and curiosity getting the best of her, she decided to take a walk past 7 Schönholzer, a foolish but irresistible urge. None of the guards stopped her—a young woman out for a stroll on a sunny summer day seemed innocent enough—and she managed to sneak a peek at the narrow lobby of the tenement as she passed. Then she went back to the church and prayed some more.

Up in the NBC apartment overlooking Bernauer, Harry Thoess started filming across the border to Schönholzer Strasse, capturing children playing, residents chatting, and VoPos patrolling on the sidewalk in front of #7. Two sheets, one white and one red, were near at hand to signal the couriers later that day. According to plans, when Peter and Klaus Dehmel got word that the first refugees might be arriving across the Wall, they would head to the mouth of the tunnel in the factory basement, and get ready to document them climbing the wooden ladder to freedom. Thoess, up in the apartment perch, would film any action on the street if something went wrong and police or Stasi suddenly rushed to the tenement. But that was still a couple of hours off.

IN late afternoon, the four-man breakthrough crew (Hasso, the two Joachims, and Gigi) gathered in the basement of the factory. To keep the noise level down at the end of the tunnel, they switched off the air circulation system. Joachim Rudolph brought with him a blue workman's coat to wear later as a kind of disguise. The tunnelers knew that in the East, by order of the state, all front doors of buildings near the border had to be locked at eight each evening. Tenants could open the door with a key, but refugees arriving after that hour would not be able to enter the tenement's lobby without knocking—which, for the tunnelers, must be avoided at all costs. Rudolph, who had some training as a locksmith, planned to dart out from the cellar and pick the lock from the inside to keep the door open after eight o'clock. Any time a tenant arrived and relocked the door, he would have to repeat this step.

What to do if a tenant decided to check on something in the basement and discovered the visitors? The intruder, like it or not, would be forced to crawl through the tunnel to the West, at gunpoint. Unfortunate, but the tunnelers were focused on the dozens eager to escape that evening.

The last member of the breakthrough team to arrive was Christian Bahner, the excitable son of the tunnelers' lumber donor. Young Bahner made quite an entrance: dressed like a gunslinger from the Wild West for this trip into the East, complete with cowboy hat, a Western shirt, boots, a gun belt, and two holsters—with pistols in them. All that was missing was a John Wayne drawl. Organizers huddled and decided there was no way this kid was going anywhere near East Berlin that day, even if he took off the gun belt, no matter what they had promised his father. He could stay in the basement in the West and cool his heels—and only if he handed over all of his ammunition.

Peter and Klaus Dehmel, meanwhile, had arrived but after setting up a few lights, could do little but wait.

Imagine the shock and alarm when diggers unaware this was escape night—and still totally in the dark about the NBC deal—came upon the lights and camera when they arrived around five o'clock. Even Uli Pfeifer and Joachim Neumann, key engineers on the project almost from the start, did not know that NBC had been filming inside the tunnel for nearly three months. Like the others learning about the NBC link, they were first angry, then deeply concerned, worrying that every additional person who knew about the escape represented one additional risk, even at this late date. And if the faces of tunnelers and escapees were shown on TV, allowing the Stasi to identify them, this might cause trouble for friends and relatives back in the East. Neumann, with most of his family still across the Wall, felt especially unsettled.

Some of the diggers challenged the NBC crew directly and demanded that they exit. Others confronted Spina and Sesta, who explained that security had been stringent, adding that the end result would be an incredible filmed document of their toil and sacrifice. The arguments went on for several minutes, putting the breakthrough behind schedule with refugees already on the way. Finally the Italians pleaded: Put your complaints and fears out of your minds for the next several hours. Everyone has a job to do. Grudgingly or not, each man accepted his assignment, which in most cases meant waiting in the tunnel to assist (or calm) the crawling refugees, including at least one elderly woman and two or more toddlers.

One of the tunnel leaders asked Uli Pfeifer whether he would mind taking a small camera to the East Berlin side of the tunnel—and filming the breakthrough site for a few minutes, for NBC. Pfeifer was shocked. "I'm willing to risk going to the East," he replied, "but not for a TV network."

AS the four Bernauer tunnelers made their way to the East, splashing through puddles up to their ankles, it was never clearer that it was now or never. Reaching the far end, they listened for any activity in the basement above. They heard nothing. Hasso had to climb onto someone's shoulders to start knocking his way into the basement. To test the surface, he first pushed a screwdriver through. The floor was not concrete, merely pressed clay, so this was easy. When he removed the screwdriver water sprayed through the tiny hole.

Had he hit yet another water pipe? Or, worse, was the entire cellar filled with water? Tunnelers had heard rumors that the Stasi flooded basements close to the border to discourage just this kind of invasion from the West. Well, there was only one way to find out. Hasso started chipping around the small hole. If indeed the basement was flooded, the tunnel, and the tunnelers, would soon be covered in water.

Enlarging the hole, he found that the culprit was merely a small leaky pipe that made everything wet but did not pose a risk to the operation. Using a hand mirror thrust through the hole, Hasso confirmed that the cellar was unoccupied. It was safe to continue—or at least as safe as the Sendlers' cottage had seemed when he had done the same thing. Neumann thought, *Well, at least we know we are not in a backyard or under a sidewalk.* Adrenaline high, the four took turns widening the hole. A glance at the window on the far side of the cellar told them it was still sunny outside. When they thought they were done, Gigi protested—they needed to expand the hole a few inches more, as he was a bit thicker around the middle than the others.

Still hearing nothing above, Herschel climbed through the gap and into the basement. Rudolph passed him the duffel bag with the tools and weapons—two pistols and the MG 42 machine gun—before climbing through himself, followed by Neumann and Spina. Hasso shoved open a door with his shoulder. Quickly they found the steps to the door that likely led to the lobby.

One major question remained: Were they *really* inside 7 Schönholzer Strasse? Or an adjacent tenement? They knew what they had to do: climb those steps, pass through that door and venture into the lobby to see if the address was posted there—and hope none of the building's tenants was checking for mail or heading out for the evening. If they had popped up under 6 Schönholzer or 8 Schönholzer, what would they do? It was too late to notify the couriers directly, so one of the tunnelers would have to loiter inside or outside #7, off and on for hours to meet escapees or—equally risky—visit one of the bars where refugees were gathering.

Hasso, with his very rare and now thick black beard, would

be too conspicuous, so Rudolph volunteered, donning the work coat he had brought along. If someone spotted him he would claim to be an electrician working in the building (which was true enough). Stepping into the rear of the lobby, Rudolph saw that the cellar door was hidden there behind the wall on the left as one entered the building, which was a plus: You could stand at the door and not be noticed from the front. He spotted another door a few feet away leading to the backyard. The lobby was the usual rectangle with staircases on either side leading upward. At any moment in the next couple of hours a tenant might suddenly descend from either side, or even appear from the backyard.

Rudolph made a quick pass through the empty lobby and spotted a list of tenants, but nothing indicating the address. He had no choice but to open the front door and step outside—just up the street from guards at the checkpoint and in full view of any policeman on patrol. Darting out to the sidewalk and looking back he saw it: a large black "7" painted on white enamel above the tall wooden front doors. (Never was the expression "lucky seven" more apt.) The five-story building looked much the worse for wear. The stone façade was badly cracked and pockmarked, probably from bullets near the close of World War II. To the left of the door the first-floor windows were either boarded or bricked up.

Barbed wire, strung on posts, ran down the center of the street just fifteen feet away. After a quick look up and down the block to gauge what refugees would have to contend with, Rudolph retreated to the basement. Neumann phoned the good news back to the West. Someone informed "headquarters" (up in the NBC apartment) that there was no need to get word to the couriers that they had the address wrong. Across the Wall, the white sheet was now draped from a window.

By then it was 6:00 p.m., and the lobby at 7 Schönholzer began to bustle with weeknight activity. The tunnelers could hear tenants arriving home from work and chatting with friends. One of the residents laughed and whistled. Others headed for the street for a bite to eat or to shop for dinner. Another spirited group, taking advantage of the warm evening, opened the door to the backyard. Rudolph, imagining what might happen if the first wave of refugees arrived in the lobby amid this much (or really, any) tenant traffic, felt like he was surrounded by hot coals. The tunnelers gripped their weapons.

WHILE this was transpiring, the Schmidt and Moeller families had arrived separately at the bar that was within two blocks of 7 Schönholzer. No one knew the area so it took some searching. The Schmidts, with Peter's mother, had taken two S-Bahns to Alexander Platz and then walked a mile, with their daughter on foot at first, carried the rest of the way. Anita and her husband had caught a tram, then pushed their baby carriage for the final blocks. All had been told to wait for a woman to enter the bar with a folded copy of the *Bild Zeitung* newspaper under her arm. The courier would walk to the counter and buy a pack of matches. The Schmidts had been instructed to depart for 7 Schönholzer as soon as the woman left, the Moellers fifteen minutes later.

Blue-collar workers began trickling into the pub for a beer after their Friday shifts. In the corner, sitting at tables just a few feet apart, were some unusual patrons, the Schmidts and Moellers, feeling quite conspicuous: two young men with the look of students or professionals, and a pair of strikingly attractive, well-dressed women tending to toddlers. One of them was wearing a fancy Dior dress and high heels. In addition, there was a gray-haired older woman. They had been sitting there for quite a while, nursing their coffees, as if waiting for someone—and starting to wonder if they were at the wrong address.

Anita, glancing at the Schmidts, whispered to her husband, "Those must be other refugees over there," but did not step over to chat. Instead she got up to take her daughter for a walk outside and visit the playground at the end of the block. Across a vacant lot she could see the death strip, the guard towers, the Wall. Spotting the barbed wire and guard shacks only magnified her fears. Back at the bar, the escapees faced confusion. A man, sitting alone, kept showing other patrons his copy of a horse-racing newspaper. Was this a signal to the refugees they had not been told about?

The infants, each less than eighteen months old, were behaving remarkably well despite soiled diapers. Annett's had leaked through to her mother's dress, but Eveline was afraid to draw even more attention by changing the diaper. An hour passed. The two mothers cast a knowing glance at each other from time to time, as they handed their kids yet another cup of water or apple juice. As seven o'clock approached Anita took Astrid out for another walk. Eveline feared that by now someone in the pub must have surely tipped off the Stasi—she felt like "a rabbit before the snake."

Then a slender young woman walked through the door, her hair wrapped in a scarf. Under one arm she carried a newspaper with the large *B* and *Z* of *Bild Zeitung* visible from yards away. Ellen Schau felt acutely self-conscious. Not only had she never done anything like this before, but as an unaccompanied woman entering a pub in East Berlin she was certain to attract maximum attention. Still, she walked straight to the bar and loudly purchased a box of matches. Glancing around the pub, it was easy for her to identify the refugees. They were the ones with stony faces and eyes glued to her every move.

Trying to act casual, the Schmidts packed up their loose items and their child, and slowly slipped past Hans-Georg Moeller on their way to the door. Out on the sidewalk, heading for 7 Schönholzer, they split up, with Peter and his mother going in one direction, Eveline and infant in the other. The evening sun cast long shadows. Hoping she had the right directions, Eveline concentrated on finding the street sign for the turn to Schönholzer. For someone from the suburbs this was confusing terrain, dominated by row after row of shabby tenements. *Keep calm*, she kept telling herself. Ellen Schau, who had left the pub and was now settling her nerves at the playground, watched Eveline walking past and silently wished her well.

Along the way Eveline passed Anita and Astrid, who were returning to the bar after their latest stroll. Eveline whispered to Anita, *Auf Wiedersehen*. See you later.

AS it happened, the two groups of Schmidts arrived at 7 Schönholzer at the exact same time. They opened the door and entered the lobby, which was vacant. *So far so good*, Eveline thought. *Now what?* They spotted a door at the end of the lobby and Eveline pushed it open—and saw only a backyard. *Nothing's here. Wrong building?* Turning around she spotted a second door in a wall facing away from the street.

Behind that door, on the steps to the basement, the four tunnelers heard new activity in the lobby. They waited, pulses racing. Forgetting to provide the password, *Potemkin*, Eveline opened the door . . . and saw several young men with weapons standing there. The one with the black beard pointed a pistol in her direction. Then Gigi Spina stepped forward to embrace his friend Peter Schmidt.

There was no time for more hugs or conversation. Carrying her baby, Eveline was guided down the steps to the corner of the basement, her husband and mother-in-law just behind. The tunnelers had rigged a lamp just inside the hole, which shone brightly, guiding the way. Almost as if in a trance, and with no alternative but to trust the next stranger, Eveline passed her child to a pair of hands in the tunnel and then, in her new dress and nylons, climbed in herself. After crawling for a spell she lost her shoes, but kept moving, past the helpers, overhead lights illuminating the chamber.

Back at the door to the lobby, Joachim Rudolph reflected: after the months of digging, the blisters, the electric shocks, the water leaks, the Stürmer scare, the constant threat of a cave-in, and his near-death experience at Kiefholz Strasse, even if no one else comes, *it will still have been worth it.*

·   ·   ·

THERE were few signs of historic drama at the Western end of the tunnel, despite the bright TV lights already trained on the scene. Tunnelers assigned to welcoming refugees to the West had wandered away from the hole in the concrete, awaiting word that the operation was under way. When the Schmidts slid into the tunnel over in the East, that news, shouted along the cavern by members of the escort team, barely reached the West. Hearing it faintly, Peter Dehmel moved quickly to angle his NBC camera to face directly into the hole, where the wooden ladder was tilted to the right at a 60-degree angle from the tunnel to the cellar floor.

The first sign of life: A woman's pocketbook placed on a ledge to the left of the ladder. Then a disheveled crown of hair, soon revealed as belonging to a young woman in a dark dress. After crawling on all fours through puddles and dirt for the past few minutes, she was now struggling to navigate the fifteen steps of the ladder. As she neared the top she turned toward the bright TV lights, her startled gaze meeting the camera. None of the tunnelers were present so Klaus Dehmel rushed to help her surmount the last few steps. (So much for Reuven Frank's order that NBC staffers remain passive observers.) Her dress was soaked with water and mud, her feet bare. NBC's lighting man guided her, near collapse, to a bench along the side of the nearby wall.

Overcome with emotion, Eveline heard what she experienced as a high-pitched sound—and briefly fainted, still upright on the bench. Regaining her bearings, she peered worriedly toward the hole. Another figure appeared on the ladder. It was Mimmo carrying her daughter, still not crying at all. Reaching the cellar floor, Mimmo passed the infant to Klaus Dehmel and bent over to embrace Eveline. Then Mimmo kissed Annett's hands, as Eveline held the child to her chest.

A half minute later, an elderly woman with badly askew gray hair climbed the ladder. When her son, Peter Schmidt, reached the cellar, he nearly hoisted Mimmo off the floor in an embrace. The four Schmidts: in the West, at last.

. . .

BACK in the East, Anita Moeller was looking for a sheet hanging from a high window directly across the Wall. She was relieved when she spotted it; more so, when she saw that it was white. Unlike the previous month, there would be no turning back. This could be a good, or bad, thing, depending on what happened next.

By now her nerves were almost out of control, fueled by copious amounts of coffee, as she passed the few pedestrians on Schönholzer. Astrid remained docile in her carriage despite her sodden diaper. The Moellers found #7 and entered the lobby. Hasso, knowing his sister was next in line, was standing just inside the cellar door. Given his black beard, Anita did not recognize him at first; when she did, they hardly spoke and embraced only for a moment. She was shocked to find him in the East at all, given his prison record and the certainty of a long sentence, perhaps even death, if arrested. Hasso said, "Go, go," and practically pushed Anita and husband toward the tunnel. They left the baby carriage in the lobby.

At the mouth of the tunnel Neumann instructed Anita to raise her hands and slide in. Hans-Georg passed Astrid to an unseen set of hands. As she crawled through large puddles and over the abrasive steel rail, still dressed in high heels, Anita complained to one of her guides, "Are you crazy, you let Hasso come to the East—somebody else should be there!" At least her dreaded claustrophobia did not kick in. On the other hand, she realized that her Dior wedding dress and nylons were getting shredded.

Emerging at the other end, knees cut and bleeding, Anita was surprised to see the bright lights. She had worked in TV in Dresden and knew that the sound she heard was the whir of a camera, and thought, *What is happening here? It's like in the East—everyone watching you*. And then she realized: *They are shooting a movie*. This couldn't be true. Her own brother had not warned her. She noticed something else odd: soggy East German currency clipped to a clothesline, hung out to dry.

Handed her daughter by a digger, she saw Eveline with her baby sitting by the wall and went over for a proper hello. Anita, unlike Eveline, had managed to retain her white dress shoes, but they shared this: the desire—no, the urgent need—to slip into the next room and attend to their kids' diapers (followed by the Dehmels with their lights and camera). Eveline still had an extra diaper in her purse but Anita had to borrow a cardigan from one of the diggers to encase Astrid from toes to chin. If either child let out a single wail, no one heard it. Then the two women took turns washing off their own legs, and the remains of nylons, with water from a basin. Eveline took Annett in her arms and felt calm enough now to smile for the camera.

Minutes later, Wolf Schroedter ushered this first wave of refugees upstairs and out to the VW van, for delivery to their lodgings for that night. In Berlin. West Berlin.

IT was now well after seven. Dusk was approaching. Above ground, Ellen Schau was nervously completing her rounds, a cigarette constantly in hand. Her third and final stop was at yet another bar near Schönholzer Strasse. The color of the sheet hanging from the NBC apartment remained white. There was just one problem: the signal in this pub was supposed to be a young woman (Schau) briskly entering, taking a seat at a table, and ordering a coffee. But this pub was not serving coffee tonight, Ellen discovered. Deliveries had been canceled. Ellen cursed the inevitable GDR shortages as she pondered what to do.

First she tried raising her voice to make sure that any refugees there would know she was trying to send a signal. "Coffee! You have no *coffee*? Why do you have no coffee?!?" The waiter appeared baffled. Not sure this was working, and realizing she sounded like a crazy person, Ellen loudly ordered something else beginning with *C* that she guessed they did have: cognac. The waiter soon returned to slap a small glass in front of her. Ellen figured she better slug it down to calm his suspicions, though she had never in her life tasted a drop of alcohol. One might say that in a few short hours she had become a highly professional courier. It seemed to work, as she noticed a few patrons stirring as if to leave.

Stepping outside, she noticed that the white sheet still hung from the fourth floor window facing East Berlin across the Wall. She wondered if her boyfriend Mimmo was up there, safe, or in the tunnel, at risk.

Duties done, Ellen grabbed a taxi that would take her to the hectic Friedrich Strasse checkpoint. Fearing that she might be searched at the border, and her stash of money deemed suspicious, she gave the cabbie a generous tip. Sure enough, at the checkpoint she was taken aside by a female police officer and strip-searched. Perhaps the Stasi knew about the escape plan after all? But memorizing routes and addresses and getting rid of most of that money had paid off: she was soon released. Ellen had engineered her own escape. Reaching the Zoo station in the West, she suddenly felt weak in the knees and nearly fainted. It finally hit her what she had done, what she had survived.

INGE Stürmer had decided earlier that week to stop listening for the knock on the door that might summon her to the West. Klaus Brunner, a well-dressed friend of her husband, had recently visited to tell her that an escape attempt was likely that month but not for at least a few weeks. So on September 14, she decided to relax and cook a rump steak dinner. She invited her aunt and uncle, as well as Doris Gerlach, a young woman she had met in Moisdorf prison. Like her, Doris had been pregnant while in prison and gave birth behind bars. They became fast friends. Also like Inge, she had an older child, too.

On this day, as Inge was serving coffee after dinner, a man on a motorbike suddenly pulled up in front of the house and a woman on the back jumped off. Inge was at the door before the woman rang the bell. The escape was on for that very night, the visitor said, in a Swedish accent, and here's where she should go. Inge would have to contact another escapee named Karin on the way. How could Inge be sure this was not a trap? The woman showed her a photo of her daughter on which Claus Stürmer had written something to the effect of *Keep a stiff upper lip.* Inge rushed to ask her aunt and uncle to leave, without explanation.

"Oh, no, not again," her aunt said, recalling the previous escape attempt that sent Inge to prison. Inge's uncle said, "Give us all your letters," knowing it might be best to burn them now.

Inge told her friend Doris what was afoot. Doris affirmed that she, too, wanted to flee. "What do I have to lose?" Doris asked, then left for home to pack a bag and retrieve her two kids. Inge promised to meet her there in an hour.

Within minutes Inge managed to organize her two children, pull on a plain checkered overcoat, and head for Doris's house, which was near the border. But—horrible luck!—the local tram was out of service, so Inge had no choice but to walk the tram route, past fifteen stops, pushing a baby carriage. Already late, she left her children with Doris and set off to find Karin. When that was accomplished, they took a taxi to Doris's apartment. Inge pulled a bunch of flowers out of a vase and wrapped them in newspaper to make it look like they were going to a party or family gathering on Schönholzer Strasse, in case they were stopped by guards. Then they hit the street: three women, two of them pushing baby strollers, and a pair of young girls. It was now almost dark. Who knew if the tunnel was still open for egress and the tunnelers still on hand.

HASSO Herschel, anxious to return to the West for a more intimate reunion with his sister, remained at his post just inside the door to the basement at 7 Schönholzer. He had a list of refugees and the order they were expected to arrive, which he had carefully plotted with the couriers. So far the list had proved accurate. Next up: a pair of women, one about thirty, the other around fifty.

When Hasso heard a knock and the password and opened the door, he found to his surprise not two women but one woman and one man. The latter was dressed in a leather coat and hat—typical Stasi wear—with both his hands in his pockets. Herschel, who already had a gun in his right hand, pointed it at the man and exclaimed, "Hands up!" When the man looked more alarmed than compliant, Hasso tried to fire off a round—but his finger hit the curved metal in front of the trigger. Before he could correct that, he noticed, as the man retreated in fear, that there was, as scheduled, a second woman behind him. Perhaps he was a husband or brother not accounted for on the official exit list?

Quickly Hasso pushed the three East Germans through the door and toward the tunnel, now fairly confident that they were harmless. He practically shook with relief. His inexperience with firearms had kept him from becoming a murderer, not to mention creating a blast that surely would have caused tenants and border guards to rush to the scene. This would have forced Hasso and the others to repeat their Kiefholz experience, madly scrambling through the tunnel in their own escape, and ending any chance for anyone else in the East to flee that night.

A few minutes later, with only half the expected twenty-five or so refugees having come and gone, Herschel and Spina left for the West, promising to send replacements. Tenant traffic in the lobby at 7 Schönholzer had pretty much ceased. The building's caretaker entered the lobby, as she did every night at precisely eight o'clock, to lock the front door. Now Rudolph would have to sneak into the lobby to unlock it. He had to wonder if his blue workman's shirt would save him if a tenant spotted him and asked what the hell he was doing there, especially with lock-picking instruments in his hands.

An hour passed. The two Joachims—thin, sandy-haired Rudolph and stocky, dark-haired Neumann—were still waiting for reinforcements or relief. No one back in the West had chosen to join them and they could not rouse anyone via their military phone. Hasso was probably out celebrating with his sister, they figured; the two Italians, with Peter Schmidt. So there they stood on the basement steps, on their own, armed but still

in grave danger. They would have liked the added security another colleague might provide, but they trusted each other not to
panic—and there was no telling who might get sent over from
the West (god forbid, not the "cowboy").

Every so often a tenant returning from work or dinner would
unlock and then relock the front door. Rudolph would have to
leave the relative safety of the basement and scurry out to the
lobby with his tools (and a pistol in his pocket), hoping no tenant
came down the steps from upstairs, and swiftly unlock the door.
Neumann, carrying the MG 42 at his chest, would "cover him"
from back at the basement door, like in an old Cagney movie,
listening for any trouble.

Refugees kept arriving in twos and threes for a while, quietly finding the cellar door and whispering the password. Neumann, as he'd done all evening, led them to the mouth of the
tunnel and into it, without a word, just pointing when necessary.
Despite their high state of anxiety, the refugees responded passively, almost as if they were on autopilot. He'd say, *Sit down,*
they sat down; *raise your arms,* and up they went; *now slide,* and
they were gone. As an engineer, Neumann thought they acted
almost like mechanical toys: silent, no pausing, no complaints—
even when they saw how dark the tunnel was, and how wet. For
all they knew, he was leading them like sheep to the slaughter,
or at least into the arms of the Stasi.

Actually, they were all making it safely to the West in an
orderly fashion. Uli Pfeifer, assisting them somewhere in the
middle of the tunnel, marveled at their composure. They just
kept splashing past. At the other end they would climb the ladder, under the gaze of the NBC camera, and then board the old
reliable VW van, heading for temporary quarters in a dorm, at a
friend's apartment, or at the Marienfelde refugee center.

At last the refugees on the master list had all arrived and
disappeared into the tunnel. Only those linked to Claus Stürmer
had not arrived, and they were late. No messenger arrived from
the West telling the two Joachims the operation was over. All
they could do was wait. Water in the tunnel, meanwhile, kept

rising. If the late arrivals did show up, they might be the last to ever escape through this passage. Neumann had hoped to bring one or more of his friends through the next day. Now the chances for a second act seemed to be dwindling.

Then three young women with four children in tow—a veritable crowd—arrived at their door and were ushered to their escape path like the twenty before them. Claus Stürmer, in the basement at the other end of the cavern, had just about given up hope that his family would make it through this night. A report crackled through the radio: Rudolph and Neumann were sending through a few women and children! Stürmer disappeared down the ladder into the tunnel, almost too anxious to hope.

At the other end of the cavern, his wife was whispering, "Don't be afraid, we're going to see Papa now," to their daughter Kerstin, who cried as they entered the dimly lit hole in the East. Inge crawled like a madwoman through the tunnel in her checkered coat, scraping and severely bruising her knees on the steel rail. When she got to the other end she did not recognize Claus, whose unshaven face was badly soiled, even after they briefly embraced at the bottom of the ladder. Inge, clueless, asked him, "Could you take my baby?"

Claus stepped all the way down to take his wailing son—whom he had never met—in his arms. From above Peter Dehmel, still on the job, captured every dramatic moment on film. Inge finally recognized her husband. A minute later other tunnelers pounded Claus on the back and hugged him. Inge thought it looked like he had just scored the winning goal in a soccer match. His wild tale the month before about only trying to get his loved ones to the West had actually panned out. Everyone, including the tunnelers, had made it to this climax, safe and sound. This time: no betrayal, no Stasi, no arrests.

Not waiting for formal notice of the end of the mission, the two Joachims finally prepared to abandon their post in the East, knowing that if they didn't leave soon they might have to swim to the West through their sturdy but leaky tunnel. Before departing, they decided that the baby buggies that had been left in the lobby might suggest to tenants that something illegal had transpired right under their noses, so the tunnelers dashed out and wheeled them to the cellar, in case the operation somehow continued for another day.

Mimmo Sesta and Manfred Krebs then crawled in the opposite direction, inspecting the state of the tunnel along the way—and taking a quick, celebratory look at the soon-to-be-famous cellar and lobby of 7 Schönholzer Strasse.

AT the NBC office, Reuven Frank was growing frantic. All day he and his editor had been watching the remarkable raw footage shot weeks and months earlier by the Dehmel brothers. He couldn't have been more excited, or anxious. Piers Anderton, who was with him most of the day, had told him the escape mission, set for late afternoon, would likely be over by eight o'clock. Since then, not a word. Frank had expected the Dehmels to arrive at the office by nine but that had not happened. He ordered an expensive Chinese meal delivered, but was too jazzed to eat it. Had the escape been compromised, as at Kiefholz Strasse? Finally he decided to go out and take a look around.

Not wishing to draw attention, he requested a nondescript car and asked one of Gary Stindt's assistants to drive him past the swizzle stick factory without slowing down. He didn't see any unusual police activity in the West nor, peering over the Wall, in the East. If there had been a major bust there would have been some sort of feverish activity or flashing lights, he assured himself, as he returned to the office, to wait some more.

At 2:00 a.m. the Dehmels arrived. They had refused to leave the basement until the mission was over, which had meant waiting, like everyone else, for the Stürmer party. Then they had hauled their precious reels of film stock to a secure lab. Reuven Frank would have to wait until later that day to screen what they had captured, but the Dehmels' eyewitness account made it sound like a phenomenal, heroic success—and, incidentally, tremendous television.

14

Underground Film

SEPTEMBER 15–30, 1962

No rest for the weary. On the morning after the mass escape, tunnelers focused on whether to try to ferry a few more out of East Berlin before the passage was completely flooded. For some, notably Hasso Herschel and the two Italians, it was already mission accomplished. Against great odds, they had extracted friends and loved ones from the East without a single arrest or shot fired. Most of the diggers wished to help others escape in coming days but the tunnel was now seriously soaked, and the risk of collapse (not to mention discovery by a tenant or Stasi agent) even greater. So they declared victory and retreated.

Others, after a quick inspection of the tunnel, wanted to exploit it for at least one more day. They discussed this at an eleven o'clock meeting. Joachim Neumann had friends he knew would flee in a heartbeat. Some tunnelers felt it was wrong not to at least try to help fellow Germans. Sure, their tunnel now hosted a dozen inches of water in spots, but all that work sawing and inserting wood supports had produced a remarkably sturdy cavern. It was too reliable to abandon just yet. Desperate young men and women frustrated by Communist rule could handle a foot of water, even if they had to do the breaststroke to the West.

A decision was made: let's try to make it work for one more day. The daring Swedish courier was willing to enter the East

again. The plan was for her to notify five people from a list com-
piled by the diggers. Those five would invite another five, and so
on. This might easily produce a score of fresh refugees. Joachim
Neumann and fellow digger Rainer Haack volunteered to occupy the cellar at 7 Schönholzer that evening and guide folks
to the tunnel. Uli Pfeifer wanted to help in some way. Claus
Stürmer said he would round up scuba gear for the escapes if it
came to that. And it might come to that.

In an apartment in Bonn, meanwhile, Birgitta Anderton
picked up a ringing phone to find her husband on the line.

"We got them all out," he announced.

"What are you talking about?" she asked.

"Oh," he replied jovially, "I forgot to tell you—we've been
shooting a documentary in Berlin about East Germans fleeing
to the West in a tunnel. And we got them all out." The newsman's use of "we" was telling.

While Anderton and most of the tunnelers were trying to
wind down, Reuven Frank was gearing up for his own mission,
which essentially had just begun. Around noon at NBC's Berlin bureau, the Dehmels' stark, silent black-and-white footage
from the night before had come back from the lab and was about
to be projected onto the only "screen" at hand: a large sheet of
stained white cardboard. Frank wondered if it could live up to
the Dehmels' synopsis of ten hours earlier.

He didn't have to wait long to find out. Just a few minutes
in, the screen showed Eveline Schmidt climbing the ladder,
with Klaus Dehmel, of all people, rushing forward to help her. It
was, Frank felt, the most thrilling moment of his entire career.
Despite the weak lighting, the image quality was technically
adequate, the tension palpable, the drama off the charts. These
scenes would require little or no narration. On and on it went,
displaying the stoic arrival of escapees and their emotional reac-
tions afterward. One scene found Wolf Schroedter leading the
first arrivals out of the bowels of the basement, past the workshop where lumber was cut, and out to the VW van. Then the
camera returned to the hole in the floor for a few more arrivals,

including the tall man in the leather coat who was almost shot by Hasso Herschel. Finally, Inge Stürmer and her children arrived, ending in a heartrending close-up of Claus on the ladder holding his baby son for the first time.

Frank was floored. He rang his boss, Bill McAndrew, back in New York to tell him to forget the plan for a sixty-minute special on a wide range of Berlin escape efforts—he would need ninety minutes just to tell the tale of this one tunnel. He had no idea how to assemble, with editor Polikoff, something powerful and coherent from the 12,000 feet of footage, shot over many weeks in short bursts, with no audio and a cast of characters unknown to American audiences. The end result, however, was no longer in doubt. Frank just had to start editing.

He couldn't imagine that, for NBC, the easy part was over.

AND what of West Berlin's newest citizens? The adults among the twenty-seven who had just made it through the muddy tunnel under Bernauer Strasse knew that in coming days they would have to report to Marienfelde to formally register and meet with German (and probably American, French, and British) intelligence agents. But that was not on their minds on this Saturday. Some, exhausted but running on adrenaline, had stayed up all night chatting with old friends or family members. Others slept well into the morning. Inge Stürmer held her jumpy daughter's hand all night until they finally dozed off around daybreak. Eveline Schmidt and her daughter had moved in with Gigi Spina's girlfriend while her husband and his mother found lodging elsewhere.

When they left their temporary dwellings for a few hours, the expatriates tried to absorb the fact that they were in *West* Berlin. Most had been there before, pre-Wall, but others were first-time visitors. What did Anita Moeller do when she left the home of Uli Pfeifer's mother on day one? Along with three of her rescuers, she went to the famous KaDeWe department store to buy diapers and clothes for her daughter, who had been wearing nothing but a borrowed sweater since the night before. Inge Stürmer, meanwhile, went to a doctor for treatment of her severely cut and bruised knees. The doctor said nothing, but she felt certain he guessed how that had happened.

Elsewhere in Berlin, another young East German citizen was enjoying her first day in the West. Angelika Ligma, a brash and pretty twenty-year-old, had long despised the Wall. The cosmetics company she worked for had enrolled her in an elite program at Humboldt University, yet she longed to begin life as a young adult in the West. That summer at the Deutsches History Museum she had met a group of West German students with Girrmann Group contacts who said they might be able to help her. One morning she packed a suitcase and told her mother she was heading off to university, failing to reveal her true destination.

In her case, the chosen method for escape was hiding under the backseat of an old Opel driven by an Italian man. And it worked, as the Opel made it through the Zimmer Strasse checkpoint in the dead of night without incident (and with the girl still breathing). Crucially, as it would turn out, this happened just hours after the last refugee crawled through the Bernauer tunnel. Ligma was taken to the swanky Kurfürstendamm shopping district, where she had coffee with Detlef Girrmann himself. He said she should live at the House of the Future for a few days and rarely leave, as it was closely surveilled by Stasi agents and he didn't want anyone from an intelligence service, East or West, or in the media to discover the car-smuggling scheme before he came up with a cover story for these new arrivals.

THE plan to spring more refugees through the water-soaked tunnel was not going as well as hoped. It was another sunny day, and a Saturday, so the Swedish courier was able to find at home only two of the five East Berliners on her list. The rest were apparently outside, enjoying their day off from work or classes. Those two, in turn, were not able to find anyone eager to flee. It was no doubt tempting to stop strangers on the street and extend this once-in-a-lifetime offer, but each might have Stasi connections.

Out of the loop, Neumann and Haack waited for three hours in the basement at #7. At any time, a tenant could have decided to check on a possession stored there. None did. Only two refugees, one of them a friend of Neumann's, arrived. The courier had tried her best but, sadly, no one else from the East would be coming. Neumann and Haack tried to hide the hole in the cellar with a bag and a baby buggy, then slipped inside for the last time. The tunnel was really filling with water now. Claus Stürmer was ready with his scuba gear but there were no refugees waiting to splash to the West. Unless someone who had heard about the tunnel arrived at #7 and figured out where it was located, or a tenant stumbled upon the hole and decided to make a hasty exit to freedom (rather than alert the police), the final total of escapees would stand at twenty-nine. Or so the tunnelers thought.

TWO days after the breakthrough, more than a dozen diggers called a meeting at the swizzle stick factory to air complaints about the failure to disclose the NBC filming and NBC money until just minutes before the operation. They were mainly students who had taken shifts in the tunnel but were not involved from the beginning nor carried the load of the core group. The idea of someone making a buck off an idealistic escape project was a shocking concept for most of them, unaware that *Dicke* and others had charged fees or received money from media at other tunnels.

Joachim Rudolph, who did know about the NBC filming and received a small sum for agreeing to appear on camera, also attended the meeting. Uli Pfeifer did not. He remained angry about not being told earlier about the money and the filming, but as an old friend of Hasso Herschel he felt stuck in the middle. (Hasso had told Uli that he nearly informed him about NBC but felt "the fewer who know about it the better.")

Sitting on the floor near the middle of the room was a man

who had long harbored ambivalent feelings, emotionally and morally, about the NBC deal. Wolf Schroedter listened as others blew off steam about the secrecy, the TV money (where exactly had it gone?), and the chance that the film would show their faces or those of escapees whose friends and relatives in the East might suffer severe consequences. Schroedter harbored some of these concerns himself but was still deeply implicated in the NBC deal. NBC had given the originators $7500 up front and promised them another $5000 to share when the film was completed. Schroedter had already resolved that he would use his fee to help fund another tunnel—with his friends at the Girrmann Group, not the two Italians, whom he blamed for causing dissension because of their excessive secrecy.

After the meeting, Joachim Rudolph returned to the wet tunnel to retrieve a few hand tools. There he also found two tiny shoes kicked off by a toddler during the escape. He resolved to turn them over to the likely parents, Peter and Eveline Schmidt.

And what of Christian Bahner, the young "cowboy" who had missed his chance to break into the East with the others on escape day? He told his younger brother Thomas that *he* had swung the ax that opened the hole in the Schönholzer cellar! And that after the refugees all made it through, he was the last tunneler remaining in the East, and when VoPos arrived they chased him into the tunnel, where he had to frantically crawl back to the West.

News about the dramatic escape did not become public knowledge until September 18, four days after the event. The *New York Post* headline revealed, "Twenty-Nine Escape East Berlin by Longest Tunnel Yet." The *New York Times* announced, "Twenty-Nine East Berliners Flee Through 400-Foot Tunnel" and claimed this was the largest group to escape in a single operation since the Wall went up. The press had been alerted earlier but delayed publishing until the flooded tunnel had been declared permanently out of service. Media were now encouraged by authorities to report that the tunnel entrance in the East was on Schönholzer Strasse. Why? "Officials made it clear they were giving details of the route," the *Times* explained, "so that East

Germans who might still want to use the tunnel would avoid being trapped in the floods."

Washington Post correspondent Flora Lewis reported that this was the sixth mass escape of the year. Earlier efforts had been organized by refugees themselves, but now volunteers in the West had taken over, unhappy with the lack of official help. They had organized their own "underground railroad." Heinrich Albertz, head of the city's internal affairs department, claimed officials had known about this new tunnel for a long time, and commented, "We do not consider it an illegal act to dig a route underneath the Communist wall." He pointedly expressed "full respect" for the diggers. This was a significant statement, Lewis argued, in light of the "quiet but bitter controversy" the tunnels had generated in West Berlin. Many citizens had declined requests to use their property or to lend other support for such escapes, fearing "unpleasant" repercussions, thus embittering the student escape helpers. Albertz, whose ministry had funded early escape actions, was now attempting to "restore the feeling of solidarity among West Berliners that is essential to their existence."

How many East Germans actually came through the tunnel? Organizers had counted only twenty-nine, but the *New York Times* reported that private sources claimed that "thirty other refugees escaped through the tunnel." Is it possible they crawled to the West after the tunnelers went home Friday night, or as late as Sunday, despite the high water level? If so, who were they? The initial report on the tunnel from the Berlin Mission to Secretary of State Rusk also cited the higher total of escapees ("twenty-nine escaped in the first group and ultimately close to sixty"). The Mission had also learned that the tunnel was built by students and, like the Kiefholz operation, "monitored" by West Germany's LfV.

THE Stasi, who had been unable to find and thwart the Bernauer tunnel despite the heads-up from agent "Hardy" and others a month before, now had to pick up the pieces. One report opened, rather bitterly, "The flight from the GDR of the twenty-nine through the tunnel supposedly has really taken place." Happily, "the tunnel is unusable now due to water ingress."

Stasi intelligence on the location of the project remained weak, however, as this report stated flatly that the tunnel was located in the far north of Berlin, "presumably in the borough Reinickendorf," several miles from the actual site. The criminals, known in Stasi-speak as *republikflucht*, "have been received by the Americans and are reaching Marienfelde one by one," the report continued. Their escape plot had been "laid out" by American intel headquarters at "P9" and now all the refugees must report there. (In fact, the Americans had already flown Peter Schmidt's mother to Frankfurt for a full debriefing about her years working at a Soviet commandant's office. They also summoned Anita Moeller and her husband.)

Siegfried Uhse had not managed to expose the Bernauer tunnel, but after winning his medal and cash bonus for the Kiefholz bust he hardly rested on his laurels. The Girrmann Group had decided it could no longer use him as a courier to the East but they trusted him more than ever, placing him in a central position in their automobile escape scheme. He was now in charge of appointing and managing couriers for the operation. Refugees would have to pay a rather modest fee of 500 DM to escape. Bodo Köhler even gave Uhse the new secret message that couriers were to use with escapees: *Nice greetings from Anuschka and Manfred, they have brought in the winter coal.* A tip from Uhse helped the Stasi arrest one of the Girrmann drivers and a fugitive passenger on September 22.

Uhse sometimes hosted his Girrmann colleagues at his apartment, even cooking a meal for Köhler at least once. Köhler found the flat quiet, safe, and comfortable, but suggested that Uhse should maybe create a kitchen nook. When Köhler talked about his own apartment, he said its location near police and military hubs made him feel secure, but he worried that Detlef Girrmann lived with much less protection elsewhere. So it would be easy, Uhse informed his MfS handler, for a Stasi operative to "hit him [Girrmann] in the head."

. . .

ONE East German who had recently crossed to the West hidden in an Opel was still hiding out at the House of the Future, but the time had come for Angelika Ligma to report to Marienfelde. Girrmann and Köhler knew Western intel agents there would grill her. To avoid unwanted inquiries about the car-smuggling operation, they decided to give her a cover story. They even made her sign a document promising never to reveal the true nature of her escape.

The cover story: She had made it to the West via the Bernauer tunnel, not stashed away in an old automobile. Ligma would have to be prepped on exactly what to say, however, because the intel services had already chatted with refugees who actually had used the NBC tunnel. Among the activists she met was a bearded young man named Hasso. Another veteran of the Bernauer tunnel, who was not identified, instructed her on what to tell the interrogators: She had met a woman in an East Berlin bar the day after the September 14 escape who told her the tunnel was still open, and she could flee the next night. With twenty-nine others, she had splashed through around eight o'clock but could not remember the address of either her entrance or exit point. Then with other new arrivals she was picked up by a bus to begin her new life in the West.

Ligma carried the tale to Marienfelde and maintained it through interrogations by all four of the Western intel services. The American operatives were especially demanding, asking some questions in different ways dozens of times. What did she know, for example, about the East German military? Well, nothing. (The Americans asked her to write one of her girlfriends in the GDR to secure more information.) By then, the Girrmann Group or the Americans, or both, had informed the media that fifty-nine, not twenty-nine, had escaped through the Bernauer tunnel that weekend, with the extra thirty muddling through when the tunnelers weren't watching. This quickly became the accepted number in the press.

While she was at Marienfelde, Ligma was approached by a press officer for the upcoming MGM film *Tunnel 28*. The Americans or Germans must have told the movie people about her. She repeated the fake story about her tunnel escape. The MGM publicist, impressed, said that maybe they could pull some strings and fly the attractive young woman to the United States the following month—to promote their film.

WHEN Piers Anderton made his first trip home to Bonn since the escape, he brought with him an odd item to show to his wife: a child's doll covered in dirt. It had been given to him by a tunneler, who said one of the kids who came through the passage had dropped it.

Reuven Frank had decided to spend a week or two editing his film in Berlin rather than heading straight back to the States. While news about the tunnel had surfaced, not one word had emerged publicly about NBC filming it, and he wanted to keep it that way. More than that, holing up in Berlin meant he could work elbow to elbow with Piers Anderton and the Dehmel brothers rather than seek advice from four thousand miles away. Gary Stindt rented editing equipment for Frank and Gerry Polikoff, and they went to work in a back room at the bureau.

Because the footage was silent—except for a few feet when the Dehmels captured the sounds of streetcars and footsteps to show how close the diggers were to the surface—it was easier to edit than most films. They had several types of footage to consider. There was Anderton's coverage of early escapes, which they quickly discarded, knowing they no longer needed it. They had Mimmo Sesta's "home movies" from trips to visit the Schmidts in the East, and the reenactments filmed that summer to portray the early weeks of the project before Anderton arrived. The largest pile of reels contained images captured by the Dehmels on their many visits to the tunnel, and then the spectacular night of the escape. Everything Frank desired was there, and then some, but he needed to find the proper pace and tone, the start and finish, and the best way to introduce "characters." With no interviews and virtually no audio, he had to produce something almost unheard-of in television—ninety minutes (minus commercials) of pure images and narration.

Strange as it might seem, Frank worried about concluding the film with escape night. It was the operation's undeniable climax, but he wanted a kind of epilogue showing how the tunnelers and escapees reacted in the days following. They were now scattered around the city and time was short, so he ordered NBC staffers to arrange and pay for a dinner party at a local restaurant, inviting diggers and some of the people they had brought to the West. To be captured on film, of course.

Many of the key "characters" made it to the party at the Würzburger Hofbräu, dressed in their best clothes: Hasso Herschel and his sister Anita; Mimmo Sesta and fiancée/courier Ellen Schau; Gigi Spina and the Schmidts; Joachim Rudolph; Claus Stürmer; and several more. Although Hasso had vowed to grow his beard until he brought his sister under the Wall, he had not shaved it off yet. Piers Anderton joined in the steady drinking. An oompah band played in the background. But the celebratory atmosphere came and went. Some of the diggers continued to mutter about the secret filming and money payments, or complained that another round of water-pumping should have been tried before the tunnel was abandoned. This made those privy to the NBC secret self-conscious. Anita Moeller even badgered her brother, only half jokingly: "It was degrading to crawl through that tunnel and then find the Americans shooting a movie!" Hasso mumbled a reply. "It was like spying," she continued. "Maybe there are microphones in the chandeliers in this room right now."

Some of the diggers were still exhausted, burned out, or eager to get on with their lives. Others had skipped the party in protest. For the escapees, the first flush of relief had passed. Now they had to contemplate a very uncertain future, with no jobs and little or no close family on this side of the Wall. Unsteady marriages had already grown shakier. For all their complaints about the Communist system, it had met their basic medical needs and offered free child care. Food was scarce in the East, but cheap. So was housing. Now, under capitalism? Who could say. They had already noted the high prices for everyday items.

One partygoer did get into the spirit of things: after quite a few drinks, Peter Schmidt grabbed his guitar and offered a sloppy, impromptu ode called "Torna a Surriento" for his Italian saviors. For this NBC turned on the audio for once. Then Piers Anderton stepped outside for his first "stand-up" related to the tunnel, facing the camera with the entrance to the restaurant behind him. He endorsed the claim that thirty unknown East Germans had come through the tunnel after the first night, adding the detail that the final group had to swim "in water over their faces." After this success, and with the hated Wall still looming, there would be "other young men" and "other tunnels" to come, he concluded.

And that was a wrap.

A few days later, the core tunnelers threw themselves another party, with less overt tension and no NBC cameras. There was a lot of drinking and even, this time, some dancing. Just two weeks earlier, Joachim Rudolph was guiding Eveline Schmidt toward the tunnel in the basement at 7 Schönholzer Strasse. Now he was asking her to dance. If she could not manage any slick moves, well, anyone could learn the new dance craze, the twist, he advised. "Just try it," he urged, and she did, and then for many minutes more. Clearly they were enjoying each other's company.

Peter Schmidt, watching them across the way, grew melancholy, then depressed. Already the excitement of the first days in the West had worn off, and marital tensions and employment issues had returned to the fore. It made him think of migrating birds who flock together in unison, with noble purpose, on a difficult and dangerous mission, and then go their own way when they reach their destination. He left the party and stood outside, nearly in tears. Mimmo and Gigi came looking for him and tried to console him. There wasn't much evidence for Peter to process, just a few minutes of a couple on a dance floor, but he felt he was starting to lose his wife to one of the brave men who helped bring her to the West. And he was right.

OTHER escapees from the East were gaining assistance from a new quarter. Rabbits roaming the death strip were setting off trip-wire alarms, causing flares to shoot into the air or triggering small detonations as guards rushed to the scene. One night rabbits near the Tiergarten produced five separate false alarms. West Berlin police claimed that in some cases refugees used the confusion and distraction to make it across the border. Neighborhood kids referred to the flares as "our free fireworks."

Beyond the Bernauer tunnel, there were no other mass escapes at the border during September, but the number of scattered small successes, and horrible failures, continued to climb. Reports filled the logs of the West Berlin police and the U.S. military's Berlin Brigade:

September 17: Three young men escape through a cellar window at 42 Bernauer Strasse.

September 20: Several more people swim across canals to the West. In the French sector two VoPos defect to the West, taking along their police dog. Another bomb blast near the Brandenburg Gate in the East.

September 23: In an echo of the Fechter incident, one youth climbs the Wall along Bernauer Strasse to freedom while his friend, under fire, falls back—but in this case no one is wounded. Two brothers, ten and twelve, flee an orphanage and cut the barbed wire at the border. Two other young East Berliners escape by climbing the Wall.

September 26: Two more Easterners escape, one of them a guard, at a checkpoint, this time on a motorbike. Border guards shoot a young man near Nord Bahnhof station; an East Berlin ambulance arrives quickly for once, and hauls him away.

September 30: "At Gleim Str a 25-yr-old female was arrested," the Berlin Brigade log reports, "by border guards not far from the Wall. The woman was waving at someone in West Berlin."

After two East German civilians dressed in fake U.S. Army uniforms crossed at Checkpoint Charlie into West Berlin, Charles Hulick at the Mission cabled Dean Rusk: "U.S. press officers have been informed so that efforts can be made to keep story out of the press should latter hear of it." Also this month, in a move that was likely to have strong repercussions, West Germany for the first time ransomed detainees in the East—twenty prisoners and twenty children. They paid with railway cars full of fertilizer. East German attorney Wolfgang Vogel (of Francis Gary Powers–Rudolf Abel spy-swap fame) was at the center of the promising initiative.

AS McGeorge Bundy prepared to leave the White House for a week in Europe he wrote President Kennedy a memo on what he hoped to accomplish. For his stop in Bonn he aimed to "show by example that White House eggheads are a lot tougher than the Germans." For Berlin, there were three goals: "To go through the regular Berlin ritual. To try and see if anything can be done about Berlin [press] reporting. To convey to General Watson in the gentlest possible way that we hope there will not be another Fechter case."

Before packing his bags, Bundy also wrote to Henry Kissinger at Harvard notifying him that the President wished to sever his consultant status. Kissinger, he advised, had tried to walk "a careful line" in claiming that in his public statements he was not speaking on behalf of the President, but reporters questioned this. So it was time for a "friendly parting." Of course, the President still wanted to ask Kissinger for advice on an informal basis.

In Berlin, Bundy attended a press conference at City Hall with Mayor Willy Brandt, who was about to leave for America. President Kennedy was absorbed in a domestic crisis—violence seemed certain at the University of Mississippi after the state prevented James Meredith from registering as a student at the segregated school—but Bundy assured the mayor that JFK would like to see him near the end of his trip. Privately, some U.S. diplomats were critical of Brandt. At the U.S. Mission, Hulick fired off a pair of lengthy cables to Dean Rusk on the mayor's "intricate balancing act"—standing with those protesting the Wall, while attempting to lower expectations that anything would actually be done about the barrier in the foreseeable future.

On September 28 the CIA circulated a classified report on "West Berlin Morale." Emotions had apparently "calmed down" since the post-Fechter riots, but the "popular mood remains volatile . . . and further eruptions might be set off by new sensational refugee incidents or additional Communist successes. A basic source of instability stems from the refusal of West Berliners to accept the division of the city as final."

BY the end of the month, after nearly two weeks of editing and working on a script with Piers Anderton, Reuven Frank was ready to return to his Rockefeller Center office in New York. The program would not be ready to air for another few weeks, but Frank was delighted with the "rough draft." It was just a matter of sharpening the pace and adding music. Anderton's narration needed to be spare, to let the drama carry the story. Anti-Communist rhetoric must be kept to a minimum.

Frank, a proponent of pictures over words, was happy to take up this challenge. A classical music buff, he was aiming to produce a kind of symphony of images, following the tunnel project from its first days to its riveting conclusion, with a tight script and set to a haunting score. In classical music terms, this meant opening with theme and exposition and then on to development, climax, and coda. One of his models was Pare Lorentz's classic 1938 documentary *The River*, financed by the New Deal's Farm Security Administration, which traced the inexorable course of the mighty Mississippi from its source to the Gulf of Mexico, across varied terrain and obstacles, with a score by famed composer Virgil Thomson. Frank saw his role primarily as a storyteller, not a "newsman," working in the confined space of a small TV screen. This story had to jump or, at least, *move*.

"The highest power of television journalism is not in the transmission of information," Frank once advised staffers, "but in the transmission of experience. . . . Every news story should, without sacrifice of probity or responsibility, display the attributes of fiction, of drama. It should have structure and conflict, problem and denouement, rising and falling action, a beginning, a middle, and an end. These are not only the essentials of drama; they are the essentials of narrative. We are in the business of narrative because we are in the business of communication." For the tunnel film, Frank had made an unorthodox decision: never let the tension flag but at times let the footage of the dirty work underground—digging, sawing logs, pumping out water, and so on—unfold at almost excruciating length. It might be repetitive, it might bore some viewers (including his bosses), but he wanted Americans to not only see but *feel* the tight, dank, dangerous conditions, perhaps even grow mentally weary, just as the tunnelers on the screen displayed physical exhaustion. This was a gamble for prime-time television.

Nearly two weeks after the tunnel escape was first reported in the press, the NBC coverage remained under wraps. American officials also seemed to be in the dark—which pleased Frank, given how quickly the State Department had learned about Daniel Schorr's plan to film the Kiefholz escape back in August. Frank had to wonder how State would respond when they did find out about it, in light of their direct warnings barring any tunnel coverage during their meeting with NBC's McAndrew in August.

Now the current cut of the film and the raw footage were

packed up in Berlin. Afraid to check their cargo for the Pan Am flight to New York, Frank and Polikoff carried it on board in hand luggage, as Piers Anderton bade farewell on the tarmac. Frank stored it behind their seats in the last row of first class, against the divider. Then a West German official walked to the front and asked the NBC team if they would mind switching seats. Mayor Brandt was on the same flight—could he take this row with his assistants? Frank and Polikoff agreed. Brandt and his party would babysit the tunnel film all the way across the Atlantic.

Frank retrieved the footage as soon as they landed in New York. He was eager to get home after the long sojourn abroad. Instead there was a note waiting for him at the airport to call his boss, Bill McAndrew. Could he fly to Pittsburgh in two days to meet with the sponsor of many of NBC's prime-time documentaries, Gulf Oil? Frank had to wonder if the content of this particular special would make advertisers as skittish as the State Department, already putting its broadcast the following month in jeopardy.

15
Threats

Within fifteen minutes of the meeting in Pittsburgh, Gulf Oil executives agreed to fully sponsor NBC's ninety-minute Berlin tunnel film. Reuven Frank felt immensely relieved. On another front, however, trouble was brewing: a reporter for *Time* had found out about the program from unspecified sources. The magazine broke the news in its first October issue, which featured Pope John XXIII on the cover. An article with no byline entitled "Tunnels Inc." explored the phenomenon of escape helpers in Berlin who received some sort of fee for their toils, notably the "musclebound ex-butcher nicknamed *Der Dicke* (The Fat Boy)." *Time* revealed that *Dicke* had no hand in the September 14 tunnel; instead, NBC had backed it in return for film rights. Rumor had it that "financing was arranged through intermediaries, two Italians and a German, who paid out what was necessary for equipment and supplies and then pocketed the remainder." In *Time*'s view, "[s]uch chicanery tarnishes somewhat the difficult and dangerous work of the idealistic diggers."

Other reports disclosed that the NBC program would air on October 31, with narration by Piers Anderton, who had flown to New York. Its working title was *The Tunnel.* NBC declined to confirm any of this. The network's secret had been exposed when journalists visited the swizzle stick factory and saw discarded boxes of DuPont black-and-white film. Knowing the other American networks had switched brands, one correspondent shouted: "NBC was here!"

With the news out, Anderton from New York wrote to his wife about "the only disagreeable development of our TV program," lamenting that it was "still supposed to be secret." A press conference to announce it had been scheduled, he added, but the *New York Times* and others "found out ahead of time." Anderton also related, "I've been writing the script—a very simple one—no poetic phrases needed—but we are still editing the film, and there is one more sound sequence on which I must appear." Gulf Oil, he said, was on the hook for $265,000 as the film's exclusive sponsor.

NBC had bigger concerns, such as viewer response to young Johnny Carson taking over from Jack Paar as host of its *Tonight Show* that week. Reaction to the *Time* scoop was swift from other quarters, however. Blair Clark, the CBS news manager (and JFK friend) who had axed Daniel Schorr's film, brought the *Time* article to the State Department's attention. This inspired a remarkable October 5 cable to the Bonn embassy and Berlin Mission from Under Secretary of State George Ball. He explained that Clark "has justifiably asked whether his excellent cooperation in suppressing"—a revealing word choice—"CBS effort on earlier tunnel project has in effect left CBS out in cold." In view of Clark's "previous cooperation," he added, "Department feels obliged give him all available information" relating to the NBC film. He didn't say what that might amount to, but the offer was likely unprecedented.

Ball also asked the embassy to find out whether Anderton's filming was "carried out with US knowledge and approval? Was he asked to desist or was this an enterprise unknown to us?" A copy of this telegram was sent to Mac Bundy and Pierre Salinger, firmly linking the White House to the NBC push-back. Salinger was a friend of Anderton, going back to their years at the *San Francisco Chronicle;* as recently as that May they had gone out drinking in Bonn. But Salinger, no less than President Kennedy, believed in journalists acting with what he called "self-restraint" on matters of national security, and didn't consider this censorship at all. But would Pierre now cut Piers some slack if State asked the White House for support in quashing another tunnel film?

The following day, Charles Hulick of the Berlin Mission fired off a cable to Secretary Rusk, also forwarded to the White House. "Reported Anderton enterprise not carried out with our knowledge or approval," he asserted. "Since we had no prior knowledge of tunnel escape . . . we were not in position to request Anderton to desist." Hulick affirmed that it was Mission policy to stay clear of tunnel activity because "advance knowledge could implicate Mission as supporting escape preparation which likely be hazardous in terms of human lives." These paragraphs followed:

> Must report that US press, TV and radio media representatives here have not accepted our reasoning and continue evidence active interest in producing documentary reports and editorials on refugee escapes. For example, Don Cook, *[New York] Herald-Tribune* correspondent, writing article for *Saturday Evening Post* on Sept. 18 [*sic*] tunnel escape. We have asked him use judgment and discretion so as not to expose and endanger individuals or write anything which could be legitimately used by Germans [to claim] that US correspondents are acting irresponsibly and thus jeopardizing future refugee escapes. . . .
>
> If and when we hear in advance of activity in which US news media may be involved, we will report and do what we can prevent dangerous involvement as we did in the case of Schorr in August. In final analysis most effective contact with US news media is that of [State] Dept with their respective headquarters in US. Must be recognized each time we intervene with correspondent here to persuade him to drop refugee enterprise we antagonize him and risk losing his cooperation.

The highest reaches of the Kennedy administration were now fully aware of the NBC filming. It was just the kind of episode that could, as State Department press officers liked to put it, "blow up" in the President's face, and in such cases Secretary Rusk and those aides would always confer with Bundy, Salinger, or Kennedy himself. The clock was ticking to an air date for *The Tunnel* at the end of the month. What were Rusk and Kennedy going to do about it?

SINCE walking away from the Kiefholz operation before its disastrous climax, Harry Seidel had hardly stood idle. Besides taking part in a cycling race or two, and treating his wife to dinner and dancing, he had returned to tunnel building. With Fritz Wagner he revisited the scene of their Pentecost tunnel success, opening up a new shaft under Heidelberger Strasse through the basement of the same Krug bar, although this one headed in a different direction. Why go to the trouble of finding a new cellar and excavating those difficult first yards when there was one ready for you? The sandy soil from the first tunnel was still stored in the basement, but the bar owner proved again a willing host, thanks to another 3000 DM from Fritz Wagner.

This time, however, the Stasi was more alert. Fairly certain that the Pentecost tunnel had started somewhere on this block, the MfS monitored the bar with a listening device and assigned an informer, code-named "Rouche," to keep an eye on the site, with star operative Siegfried Uhse also ready to help out.

The new tunnel under Heidelberger, just a few blocks from the cursed Sendler cottage, was only about eight feet below the surface. For security reasons most of the diggers had agreed to stay in the basement under the pub for up to three weeks without exiting. Fritz Wagner, meanwhile, had only lined up a dozen refugees despite the usual financial incentives. Mertens, the LfV operative, learned about this and once again asked the Girrmann Group to help, despite what happened after he invited them into the Kiefholz project.

The basement of a master tailor named Castillon, who

wanted to flee with his wife, would serve as the breakthrough point in the East. When Seidel hit the cellar wall, word went out to the Girrmann Group to rally the refugees. And who was invited to the key meeting of escape helpers at the House of the Future on the evening of Friday, October 5? Always in the right place at the right time: Siegfried Uhse, who had known nothing about the tunnel until then.

By now, about forty refugees were ready to flee. Some had been ticketed for escape earlier at the Kiefholz tunnel. Several in the new group were East German soldiers planning to defect. Seidel had invited a couple of his old cycling buddies. A professor of medicine from Leipzig was on the list, too. This might provide a nice payday for *Dicke*. Organizers at the Girrmann meeting that night outlined what was to transpire starting in a few hours, including the signals to be used for refugees, who would arrive at intervals in small groups at the tunnel entrance.

Asked to participate, Uhse said he didn't have any couriers ready, and that he was leaving for vacation the following day (a rare truthful claim by the young man). But when the meeting broke up around midnight, he was not too busy to rush to the Friedrich Strasse checkpoint and call his Stasi minder, Herr Lehmann, no doubt rousing him from sleep. When they met at the station within the hour, Uhse passed on full details about the escape, including the use of a sheet hung from a window as a signal to refugees—red for stop, white for go, just as for the Bernauer tunnel.

At daybreak, before the Stasi had a chance to mobilize, Harry Seidel broke through the basement wall and stepped into the East. Seidel found the tailor and his wife still in bed. Quickly they scrambled into the tunnel, the tailor still in his nightshirt. Harry retrieved some of the couple's belongings for them, confident that he had many hours to get everyone else through. Pausing in the pub's basement to rest and celebrate, Seidel sent over Erhard Willich, a longtime Wagner digger, and Dieter Reinhold, one of Harry's old cycling friends. They carried a submachine gun and a pistol, but—as usual with escape helpers—neither had ever fired a weapon. Taking their position at the basement door, they waited for a bell that would signal they should open it to greet a refugee.

Thinking he heard the correct signal, Willich opened the door a crack—and spotted four Stasi with machine guns. He ran to the tunnel but too late. Bullets came flying through the door, hitting him in the arm, leg, and thigh. Reinhold left his comrade and crawled back to the West. Rudi Thurow, the former East German guard, started to scramble to the East to save Willich, but Seidel shouted, "Stay here! Are you crazy?" Willich, bleeding, was captured and hauled off to a hospital. Four refugees arriving on the scene were arrested, but others, hearing gunfire, fled.

Returning home, after again barely skirting arrest or worse, Harry Seidel faced his wife's fury: "Jercha dead! Willich shot! When will you give up?"

Harry, ever hardheaded, replied, "Not until I get my mother out."

Fearing more riots, Charles Hulick cabled the news to Dean Rusk, concluding with: "U.S. and British instructing police take all necessary precautions prevent demonstrations against Soviet war memorial buses . . . and along Wall in U.S. sector where incident occurred." Siegfried Uhse left for his vacation in Switzerland with an extra 200 DM, a reward handed to him in the middle of the night by Herr Lehmann.

FOR several days NBC chose to remain silent following the reports that it had paid organizers for the right to film the Bernauer tunnel. The State Department also made no public statements. Then NBC announced it would hold a press conference to confirm its plans for the program.

On October 10, Robert Manning, State's public affairs chief, called Bill McAndrew of NBC News, expressing regret that his network not only had taken a foolish risk in filming the project but was now going ahead with a prime-time special. State would not openly attack NBC, he added, but it would reveal that back in August CBS had dropped plans to film an escape while NBC, despite being warned, plunged ahead. The message was clear: Dean Rusk did not want to be accused of suppression or censorship—but State would make things so hot for NBC that the network itself would kill the film.

At NBC's press conference in New York, Reuven Frank confirmed details about the program and the roles played by Anderton and the Dehmels, and admitted that the network had indeed paid the three organizers. However, he insisted, "[w]e were not going around recruiting tunnel builders," and he claimed the project would have been completed without the NBC money. Frank refused to say how much NBC had paid. "Not much in television terms," he said dismissively. NBC circulated a press release in which Frank described the tunnel as "not much roomier than a coffin." Harry Thoess, the release revealed, had shot as much footage above ground, about 6000 feet, as the Dehmels had captured below.

Richard S. Salant, the president of CBS News, issued a statement confirming that his network had indeed planned to film a tunnel escape back in August. He wrote pointedly: "After receiving from the State Department certain intelligence information on the United States national interest aspect of the Berlin Wall tunneling operations, CBS News stopped the preparation of its reports and has not resumed it."

Manning promptly informed Dean Rusk that NBC had gone ahead with its press conference, despite his warning of the day before. He attached the Salant statement to his memo, adding that CBS had contacted him to emphasize that "while they were sorry to be caught in a competitive situation with NBC they still accept today as thoroughly as they did before the correctness of the Department's interference in their tunnel activity and the correctness of their decision in withdrawing the project."

A *New York Times* headline the next day clearly pegged NBC as vital, not incidental, to the Bernauer project: "NBC-TV Plans Documentary on Berlin Tunnel It Helped Build." That promised nothing but trouble for the network in days ahead. One of the paper's top writers in Washington, Max Frankel, weighed in with an analysis entitled "Cold War Confusion: Washington Wonders Who's in Charge as Exploits to Harass Reds Multiply." Among other controversies, he cited CBS's feeling "professionally undone" by NBC going ahead with its tunnel. Frankel observed, "Officials are wondering what bizarre enterprise the competition will produce next."

All of this led Piers Anderton, still in New York, to write to his wife, Birgitta, back in Bonn:

> There has been much more of a turmoil than I expected. CBS is trying to get the State Department to ban [the program], and there is still a chance that [NBC president] Kintner may back down. . . . I have been too busy to get much involved in all the peripheral foolishness, but I have to go on radio and television shows to promote the program, so I am bound to get somewhat mixed up in all the stupidity.

A state-run news service in Moscow now offered its first commentary. It claimed that NBC had "hired" the diggers from the outset and that the role of refugees "was played by various rogues who received an artistic honorarium of 2,500 dollars each." Even after its support was unmasked, NBC had decided to go ahead with the faux film, which would do nothing but "strengthen the policy of provocation in Berlin." For once, some West German officials in Bonn, as well as American officials in D.C., seemed to agree with the Communist press.

SIEGFRIED Uhse was up for another award for his role in busting the latest tunnel—making that two in a row for him regarding Harry Seidel's projects. As with the Kiefholz episode, Uhse had acted swiftly when he realized he was suddenly at the center of the Heidelberger operation. His commendation came with 500 DM in cash. It noted that thanks to his "quick and courageous action, a large-scale forceful border breakthrough by a West Berliner terrorist group" had been destroyed. Young Uhse "demonstrated major devotedness, reliability, swift reaction and personal courage during the execution of his orders for the MfS."

Stasi headquarters, meanwhile, updated its handwritten log of suspected tunnel operations. Successful or abandoned projects were added as physical evidence was discovered, or when informers or those arrested spoke of them. A few were still unconfirmed. Columns noted where each tunnel was located, the date it was discovered, and when it was liquidated (if it had been). According to that ledger, for example, the Bernauer/NBC tunnel was uncovered on September 25, eleven days after the escape; "Girrmann" was falsely listed as its organizer. Elsewhere in the log the names that showed up most frequently were the Franzke brothers, Wagner, and Seidel.

Total suspected tunnel attempts through October: 137.

From far away in California, a staffer at the Stanford University radio station KZSU telephoned Joan Glenn, the Girrmann escape helper in Berlin. Glenn revealed that, since his arrest early in the year, Stanford classmate Robert Mann had provided Stasi interrogators with no information about escape operations. Glenn called this "quite a feat." She described smuggling techniques at some length, but she did not mention anything about tunneling (nor the fact that she reputedly slept with a pistol in her bed). What did she now think about that early period of passport smuggling? "We were a little blind to any danger," she admitted. "We wanted to help. I know this sounds naïve but when you are asked to help by an East German, it is impossible to say no." As for her friend, the former courier Mann: "Bob may be released by Christmas, because the propaganda value of his imprisonment is gone."

AS its airdate neared, production of *The Tunnel* was coming together nicely. A musical score had been commissioned. Frank and Anderton had completed the script. They tried, not always successfully, to eliminate anti-Communist rhetoric and references to "the Free World," but were pleased to have added what they felt was the key line in the film. That came about when

Frank screened a rough cut for another producer. After viewing several minutes of refugees climbing the ladder to the cellar of the swizzle stick factory, frazzled and muddy, their clothes
ruined and hair askew, the producer commented, "What must
they be leaving to risk this?" Frank added it verbatim to the
narration.

The network's publicity department was just gearing up
when a new threat emerged. On October 14 a front-page *New
York Times* headline announced: "Tunnelers Fight Berlin TV
Show / Deny NBC Has a Right to Present Film of Escape."
Seventeen men claiming to have worked on the project had
banded together in an attempt to keep NBC from showing the
film, or at least edit them out of it. Their spokesman was Eber-
hard Weyrauch. At fifty-eight he was probably the oldest tunnel enabler, and he was no digger. Weyrauch had raised funds
for the project, however, from the Axel Springer company and
private groups, and also had excellent contacts with West German police and U.S. intelligence. Now he and others had asked
the Berlin Senate and the West German embassy in Washington to exert pressure on NBC. The network had no right to air
the film because the filming was done "without our approval,"
Weyrauch charged. The protesters were "disgusted" that anyone
had pocketed money; they were only out to help others. In addition, airing the film might endanger the lives of the diggers or
their families in the East, as well as future tunnel operations.

NBC took this seriously enough to consider obscuring the
faces of the protesters in the film. Fortunately, few of them
appeared in the footage, and then only for scant seconds. Far
more serious was the demand by Hasso Herschel, who was not
covered by the original NBC contract but who *did* play a key
role in the narrative. Herschel, who had previously received just
1000 DM (or $250), now asked for additional pay, as well as the
same right to sell photos and film footage in Europe. Otherwise,
he declared, no images of himself or his sister and family could
appear in the film. The network quickly agreed to add Herschel's
name to the contract with the three organizers. NBC also prom-

ised to pay Rudolph and Casola, and the Schmidts and Moellers, a small sum of money, but they were still locked out of the much higher amounts likely to arrive from foreign sales. Gigi Spina and Mimmo Sesta were already selling pictures to magazines through a photo agency.

Peter Schmidt managed to get NBC to reimburse his mother for the 3000 DM she had invested in the project that spring, but he quarreled with his old friend Gigi about not sharing in subsidiary rights. His wife, Eveline, however, didn't care about the money one bit. She had other concerns. Eveline was starting to have nightmares about escape night. Following the Kiefholz tunnel fiasco, she had managed to suppress the memory of the terrifying moments when she and Peter realized that Stasi agents were observing them as they walked toward the Sendler cottage. But now the fear she had overcome on escape night at 7 Schönholzer was returning in full force. One night, sitting at a table with Mimmo and other visitors at a friend's apartment, she suddenly broke down and started screaming. For fifteen minutes, she could not stop.

AS the broadcast date drew closer, NBC still offered "no comment" on the protest by the seventeen ex-diggers. Not so U.S. officials in Germany. From Berlin, Allen Lightner cabled Dean Rusk on October 15 that the protest group might okay the NBC film if the network excised certain names or faces. This possible solution apparently displeased Lightner, as it would leave NBC's plan to air the film intact. He urged the State Department to confer with the German embassy "to coordinate possible new approach to NBC" that would go beyond those personal concerns, which might soon be settled, to one involving national security, which would not. The following day he informed Rusk that the influential German minister Heinrich Albertz had told the media that he, too, opposed the NBC project. The West Berlin Senate endorsed that view and was contacting the German embassy in Washington.

On October 16, UPI quoted an unnamed State Department spokesman declaring that the film would "complicate the
Berlin situation," and claimed that NBC made the film "after
having been asked not to do so." Now *The Tunnel* was truly
threatened—so much so that NBC executives decided to take
drastic action. They enlisted their vice president for corporate
affairs, Lester Bernstein, to board the next flight to Germany
and meet with top officials to dispel fears and plead the network's case. Bernstein, now forty-two, was the master diplomat
at NBC who had hammered out logistical details for the ground-
breaking Kennedy–Nixon TV debates in 1960. Before that he
had served as film critic for the *New York Times* and as London
correspondent for *Time*.

While this was transpiring, Egon Bahr, the aide to Willy
Brandt perhaps most sympathetic to *fluchthelfer*, rang a dorm at
TU looking for Hasso Herschel, whom he had met that spring
when tunnelers were seeking funds. Herschel, summoned to
the phone while drinking a beer, was shocked to hear Bahr on
the line. "Do me a favor," Bahr said, "and tell NBC not to show
that film! We are trying to not provoke the Soviets and keep the
situation calm. Remember, we are a four-sector city and you are
only making trouble!"

"No," Herschel replied, unhesitating, and maybe a little
tipsy. "We have so much hate for our enemies . . . it's impossible
to stop." In fact, Hasso was already plotting his next tunnel, to
be financed by his NBC loot.

Even a pro-Western newspaper in Vienna slammed NBC for
its "thoughtless, business-minded capitalism" in encouraging
East Germans "to risk lives so American families, safe at home,
can get thrills by watching [the] escape on TV. . . . Next to the
repulsiveness of the tyranny which causes people to flee is repulsiveness which pays for and films the flight."

Ratcheting up the crisis, Secretary Rusk on October 17 asked
Robert Manning to inform NBC that it should give "every consideration" to the "expressed view" of West Berlin officials on
the program. Manning contacted NBC's Bill McAndrew and
passed on the West Berlin Senate's "requests that NBC abandon

its plans to telecast it." The Senate was of the opinion that the film "will offer to the Communists propaganda material for use against West Berlin." The West German embassy endorsed this view "and urges you to drop your plans to show the film." Then, absurdly, he closed by repeating that while State continued to have "strong reservations" about any TV involvement with tunnels, the department "feels it must leave to you the question of whether the film should be televised." It sounded like he was taking cues from President Kennedy's angry April 1961 speech to newspaper publishers, when he had urged every journalist to exercise self-discipline to voluntarily "prevent unauthorized disclosures to the enemy."

McAndrew happened to be in Washington, so he stopped by Manning's office as a balmy mid-October afternoon topped 80 degrees to clarify NBC's position. The conversation was civil, even pleasant, in stark contrast to the August meeting between McAndrew and Manning's deputy. McAndrew assured Manning that NBC had taken precautions during the digging to reduce risks. The network had not "financed" the tunnel, simply shelled out money for the right to film the diggers in action—no different from previous cases when the media paid newsmakers for interviews and photos. (There was no mention that the tunnelers had desperately needed, and used, NBC money for equipment and supplies.) McAndrew also falsely claimed that NBC staffers had not actively aided the tunnelers in any way.

As for the network's film in progress, McAndrew said they would eliminate identification of anyone who did not wish to be filmed, by either cutting them from the program or obscuring their faces. But he held firm on not canceling the show, calling it one of the most compelling "human" documents ever produced for television, celebrating as it did man's unconquerable drive to be free.

CONFLICT swelled again later that day when McAndrew, Frank, and Elie Abel, NBC's State Department correspondent, met with

Secretary Rusk at 5:00 p.m. in his elegant seventh-floor office. (Rusk's secretary wrote Reuven Frank's name as "Mr. Rubin" on the appointment calendar.) Abel had a delicate relationship with President Kennedy, a man he considered a friend. After winning the 1960 election, JFK had pegged Abel, then a newspaper reporter, for the job of Robert McNamara's press officer. He declined despite a substantial increase in pay. Next he was targeted to serve as Dean Rusk's public affairs chief. Again he turned it down, recommending Robert Manning for the post. Then Abel took the NBC job.

Now he was sitting with the President's top man at State, in a meeting he had brokered. When Rusk suggested that the tunnel film might not be in the national interest, the newsman nearly lost his temper. "This is a program about human freedom!" Abel exclaimed. "Does the Department consider *that* in the 'national interest'?" The rest of the meeting proceeded along these lines, in paraphrase:

Rusk: It would have been embarrassing or worse if East German police had raided that tunnel and discovered anyone from an American TV network.

NBC: But nothing like that happened and the tunnel was a success.

Rusk: Well, when the program airs you will be tipping off the East Germans on who escaped.

NBC: This would be the first time they ever needed our help to do that. Are you saying that we should cancel the program?

Rusk: It would be improper for State to suggest such a thing. It's up to NBC.

NBC: But all the messages from State and newspaper reports add up to pressure to cancel.

Rusk: That's the furthest thing from our mind.

This was Reuven Frank's interpretation: *Rusk doesn't ever want grubby reporters getting in his way.*

FOR over a month, leading Republicans, and a few media pundits, had charged—based on little apparent evidence—that the Soviets were placing nuclear missiles in Cuba. CIA director McCone, a Republican, had privately voiced this fear to the President, who remained skeptical even as he expanded U-2 surveillance of the island. JFK was not alone in feeling that this might turn out to be just another GOP campaign issue that would evaporate after November. Berlin remained his constant concern. He had told Ted Sorensen, "If we solve the Berlin problem without a war, Cuba will look pretty small. And if there is a war, Cuba won't matter much either."

As weeks passed, however, a Soviet military buildup continued in Cuba, so Kennedy ordered more careful study, including a U-2 flight directly over the island on October 14. The following day CIA analysts in Washington studied the photos that resulted. New missiles longer than the defensive surface-to-air missiles (SAMs) seemed to have arrived on the island. They could only be medium-range, nuclear-capable SS-4s or their cousins. The President was in New York on a campaign trip, so McGeorge Bundy waited until the next morning to inform him.

When October 16 dawned, President Kennedy, back in D.C., responded to Bundy's startling news by calling the first meeting on the confirmed nuclear threat for later that morning. The missiles were not yet operational, but soon would be, with the capability to strike at least the southern half of the U.S. mainland. Already JFK was outlining four options: hit just the missile sites; blast them plus other military sites; do all that and also enact a blockade; or all of that plus invade the island. On his notepad he scribbled scraps of words: *Prepare. Berlin. Preparatory. Cuba. Preparation. Cuban uprising. prepare. nuclear.* Vice President Lyndon Johnson, more hawkish than nearly anyone else in attendance, volunteered that massive air strikes could be carried out without even informing Congress or America's allies. The meeting ended with the Pentagon ordered to study how air strikes and an invasion might proceed.

Robert Kennedy handed his notes on the meeting to the President's secretary. One read: *I now know how Tojo felt when he was planning Pearl Harbor.*

At a morning meeting the next day, George Ball argued against military action, claiming the Soviet leaders didn't really understand the gravity of their Cuba move. Ambassador Llewellyn Thompson, back from Moscow, took issue with this, stating that the Soviets' aim was to force a showdown on Berlin; General Maxwell Taylor and John McCone agreed with him. Kennedy left for a long-planned meeting with West German foreign minister Gerhard Schröder. He did not mention Cuba to him, but Kennedy had to wonder, *Are the Germans ready for a Soviet retaliatory strike or another Berlin blockade?*

With JFK away on another campaign trip, aides and military leaders discussed options, as air strikes—with or without a warning to Khrushchev, and with or without a U.S. invasion—gained wide favor. (The military projected 18,500 U.S. casualties in a normal invasion, but if nuclear weapons were used, General Taylor dryly advised, "there is no experience factor upon which to base an estimate of casualties.") George Ball responded with a strongly dissenting memo, arguing that air strikes would smack of Pearl Harbor—echoing Bobby Kennedy on this—and turn much of the "civilized world" against America. But some felt that a strong U.S. action on Cuba, rather than jeopardizing American standing in Berlin, would boost its credibility in dealing with the Soviets down the road.

Now the President had returned, and at an October 18 meeting he repeated his argument that the Soviets, in response to any military attack on Cuba, would likely "just grab Berlin." Sacrificing a few Cuban missiles for control of all of Berlin would not bother the Soviets at all. And once they took the city, "everybody would feel *we* lost Berlin, because of these missiles." Bundy cracked a small, revealing joke: "If we could trade off Berlin—and *not* have it our fault . . ."

Then came a grim exchange after JFK again said that the Soviets, if attacked in Cuba, would probably cross the Berlin border with ground troops.

McNamara: We have U.S. troops there. What do they do?

Gen. Taylor: They fight.

McNamara: They fight. I think that's perfectly clear.

President Kennedy: And they get overrun.

Robert Kennedy: Then what do we do?

Gen. Taylor: Go to general war, if it's in the interest of ours.

President Kennedy: You mean nuclear exchange. (*Then a brief pause.*)

Gen. Taylor: Guess you have to.

The President, therefore, insisted on considering "what action we take which lessens the chances of a nuclear exchange, which obviously is the final failure." Apocalypse had reared its ugly head, just in time. Henceforth the discussion—led by Rusk, McNamara, and Robert Kennedy—shifted in a more dovish direction, toward trying a blockade. The Joint Chiefs of Staff opposed this, still favoring a preemptive attack, but within minutes a consensus began to coalesce among the nonmilitary aides in the room. A blockade was best, along with a potential concession to scrap outmoded American nuclear missiles in Turkey. McNamara pleaded, belatedly, "I really think we've got to think these problems through more than we have."

At a crucial point, an unlikely player had contributed to the meeting's change in tone, appealing to Kennedy to insist on a blockade: "Essentially, Mr. President, this is a choice between limited action and unlimited action—and most of us think that it's better to start with limited action." The speaker was Roswell Gilpatric, number two man at the Pentagon, who just two months earlier JFK himself had fingered as the likely leaker in the Hanson Baldwin case. Gilpatric, seemingly forgiven—and despite his history of leaking secrets to a journalist—had been attending crucial meetings on the missile crisis and joining in the highly classified discussions. (The FBI investigation into both Baldwin and Gilpatric, meanwhile, continued.)

After the meeting broke up, JFK met with Andrei Gromyko, the Soviet foreign minister. Gromyko lied to Kennedy's face, denying the presence of offensive missiles in Cuba as the President tapped his desk, exercising supreme self-control, knowing that he had photos that proved otherwise close at hand, in the top drawer.

Near midnight, after everyone had left the Oval Office, Kennedy turned on his taping system to record his summary of this day. Although the crisis revolved around Cuba, much of his musing still concerned Berlin. He noted Bundy's warning of "a Soviet reprisal against Berlin" after any U.S. military action. Others felt that failing to take strong action on Cuba would undermine American promises to the West Germans, "divide our allies and our country." JFK concluded, "The consensus was that we should go ahead with the blockade beginning on Sunday night." If the Soviets did move on Berlin—well, the United States faced "a crunch" there in a few months, anyway. And a blockade would be a lot less likely to inspire that Soviet reaction than a military assault on Cuba.

Then he went upstairs to sleep on that decision, if he could.

16
Buried Tunnel

OCTOBER 19–31, 1962

Two days after meeting with Dean Rusk in Washington, Reuven Frank sought refuge in a New York studio to oversee the recording of his film's soundtrack, mixing Piers Anderton's dry narration with an eclectic score composed by Eddie Safranski, a well-known jazz bassist. The tone of the music, recorded by a small ensemble, shifted from tedium to tension as the images did, with an occasional comic touch. Suddenly the door opened and someone handed Frank a UPI story. A State Department spokesman had just declared that airing the tunnel show was "irresponsible" and "not in the national interest." That certainty was striking, since no one at State had even asked to screen the film. Reading the article, Frank wondered if NBC, which had stuck its neck out for so long on his project, could withstand the pressure as *The Tunnel* became an international cause célèbre.

The State spokesman was Lincoln "Link" White, a department mouthpiece since 1940. He had received two pages of "guidance" from his superiors for that day's press conference. If asked about the CBS tunnel, he was to affirm that the network's cancellation of its coverage back in August had been "most appreciated by the government." When asked about the NBC tunnel, he was to acknowledge that State had expressed the same concerns it had passed on to CBS—that their program "could well complicate the Berlin problem."

This script had been cleared at the White House by Pierre Salinger, who exercised direct control, and accepted responsibility, for statements by press secretaries in any top government department, from Post Office to Pentagon. He liked to refer to this as "a community of information officers." Face-to-face meetings with the key spokesmen were often held on Tuesdays. To improve communication with State and Defense, Salinger had installed a telephone system that allowed him to converse with both at the same time. Many times he took an important statement from one of them directly to the President for approval.

Sure enough, the first question on the NBC documentary came shortly after the press conference started at 12:35 p.m. As instructed, Link White praised CBS's response. Then he went off script, calling NBC's film "risky, irresponsible, undesirable and not in the best interests of the United States." NBC had chosen to "continue with its tunnel project" despite objections from West Berlin and West German officials. "I think our view on the question is quite clear," he declared.

A reporter asked the obvious: "Link, I'm not quite sure what is meant here when it says 'the views of the Department are quite clear.' Is that implying whether the film should be shown?" White cited unspecified risks. The reporter pressed, "In other words, the Department is against the showing of the film?"

"I don't want to go beyond what I have said today," White replied.

Another reporter asked what the "risks" were now, since the tunnel operation was over and, in fact, successful. White cited the possibility of "reprisals" for tunnelers shown on camera or their families remaining in the East. The reporter continued, well, doesn't that mean you surely want the film axed?

"I have said enough on this subject," the State Department spokesman said, growing testy. "I think it's clear what I'm saying."

Not to all. Another journalist asked what he meant by NBC's "involvement" with the tunnel. What was it exactly? White, like a modern-day Will Rogers, claimed he only knew what he read in the newspapers.

Then a mischievous reporter, citing the CIA's well-known Berlin spy tunnel from West to East in the 1950s, chimed in with "When did the United States government turn against tunneling operations?"

After the press conference mercifully came to an end, White was called out by his superiors (and probably Salinger), for going too far in his denunciation of NBC. White issued a statement to prevent a "misunderstanding." The State Department did not mean to imply that airing the film would "risk lives" of those involved in the tunneling—only that NBC's involvement was risky during the actual operation. "Additionally, it is not the intention of the department to ask NBC to refrain from showing the film . . . this is a matter for NBC to decide," he said.

This clarification only confused matters. Reuven Frank felt puzzled and depressed. Frank believed that all government officials, especially those involved in foreign policy, wished the press would just go away and let them do their work. That sentiment had to be resisted by media at every turn. Now he feared that his ultimate boss, NBC president Robert Kintner, was growing skittish. Kintner worried him. True, he was a longtime journalist, the head of the network since 1958, and a promoter of documentaries. On the other hand, he was close to Vice President Johnson. He had hired and then injected quiz-show champion Charles Van Doren into Frank's news programs (before the cheating scandal broke). Frank considered Kintner an alcoholic and a bully—stubborn, intolerant, impulsive—though also partly a genius. Kintner rarely interfered with news programs, which made his recent intervention, sending Lester Bernstein to Berlin, even more ominous. It was starting to look like NBC would at least postpone, perhaps even cancel, *The Tunnel*.

Still in New York, Piers Anderton wrote to his wife back in Bonn: "This visit to New York has certainly been no pleasure. Every day a new crisis, and today it looks like the program may not go on the air, or may go on cut to ribbons. If the latter, neither Reuven nor I would have anything to do with it. The U.S. and the West Berlin Governments are putting on heavy pressure to stop NBC from putting the show on the air, and Kintner is weakening."

Jack Gould, the influential *New York Times* television writer, did not help matters when he observed that whatever NBC's motives, "one effect has been to inject a note of distasteful commercialism and possible hazard into an area of extreme delicacy in the cold war." Gould admitted that as far as journalistic principle went, NBC could not be criticized strongly. Look at American astronauts selling their personal accounts to *Life* magazine. And if a print journalist had "bought" the tunnel and written about it, controversy would have less likely followed. But for TV, with its new "influence and impact," one could expect a backlash. As much as journalists might defend NBC's project, to the general public the "stamp of commercialism" had been applied to the tunnel, which gave the Communists an excuse to claim that "the sign of the dollar" lurked behind any escape activity in the West. Gould concluded:

> In many ways the cold war requires exercise of greater restraint than that asked in the hot war. With peace hanging by a thread it is not time for adventurous laymen to turn up in the front lines of world tension. No one knows what seemingly small provocation might be grasped as an excuse to press the proverbial button. . . . [T]he cold war is complicated enough without the introduction of video's competitive economics.

WHILE the "stamp of commercialism" surrounding the true-life Bernauer tunnel offended, among others, certain West German officials, they were warmly embracing the Hollywood version. An MGM publicity release revealed that when its *Tunnel 28* film was completed, the West German government "summoned Producer Wood to Bonn, where he showed officials the rough cut. They were so impressed they immediately asked the producer to plan for a premiere in Germany."

Another tunnel fiction earned official support in Bonn on

a smaller scale. Ernst Lemmer, the minister for all-German affairs, had secured a visa in just a few days—rather than the usual two years—for Angelika Ligma, the young woman who falsely claimed she had escaped to the West via the Bernauer tunnel. This was on behalf of MGM, which wanted to use her to promote its movie. Allen Lightner informed Dean Rusk that the American consulate in West Berlin had issued the visa to Ligma. She had been directed to MGM by Lemmer and the Berlin police chief, "who desired [to] assist" the studio's plan to use her in America "to exploit" its film. Lightner's cable ended with a warning: "Dept. may wish contact MGM or suggest German embassy do so to assure exploitation Ligma not counter-productive to Berlin cause."

Across the border, the Stasi already knew that Angelika Ligma was in West Berlin. She had sent a letter to her mother from there. Her mother informed the MfS.

BY nightfall on October 18, President Kennedy thought his executive committee, or ExComm, had reached a consensus on blockading Cuba without firing a shot in anger, but the next day this began to unravel. An early morning meeting with the Joint Chiefs, who still strongly advocated a preemptive U.S. strike, tested JFK's resolve. McGeorge Bundy had shifted to that shoot-first position overnight, and it seemed that everyone's emotions were dangerously fluid.

The President continued to play the Berlin card. Any attack on Cuba, he maintained, would give the Soviets "a clear line to take Berlin. . . . We would be regarded as the trigger-happy Americans who lost Berlin." European allies would be livid—they cared passionately about Berlin and their own security but didn't "give a damn" about Cuba. And if the Soviets moved on Berlin this would leave him "only one alternative, which is to fire nuclear weapons—which is a hell of an alternative—and begin a nuclear exchange. . . . When we recognize the importance of Berlin to Europe, and recognize the importance of our allies to us, that's what has made this thing be a dilemma for three days." Remember, he added, the "argument for the [Cuba] blockade was that what we want to do is to avoid, if we can, nuclear war."

Rarely have three words—*if we can*—signified so much.

General Curtis LeMay, the most hot-blooded hawk of the Joint Chiefs—one of his nicknames was "the Mad Bomber"—opposed a blockade on grounds that it would "lead right into war." He called it "weak" and compared it to "the appeasement at Munich." He wanted a full bombing assault as soon as possible. LeMay addressed Kennedy, "You're in a pretty bad fix at the present time."

Angry, Kennedy asked LeMay to repeat that. Then JFK replied with a laugh, "You're in there with me!"

After the meeting Kennedy, always skeptical of advice from the military, said to an aide, referring to LeMay and his ilk, "These brass hats have one great advantage in their favor. If we listen to them, and do what they want us to do, none of us will be alive later to tell them that they were wrong."

Still, the press and the public were kept in the dark. A Pentagon spokesman stated in response to a media query, "The Pentagon has no information indicating the presence of offensive weapons in Cuba." The mechanics of a possible—some believed, likely—retaliatory Soviet blockade of West Berlin gained wide private study in Washington, however. David Klein, a National Security Council assistant, submitted a top-secret memo to his boss, Mac Bundy, under the heading "The Defense of Berlin if Cuba is Blockaded." Klein bluntly advised that a Soviet blockade "would inevitably stir feeling among all Europeans that this crisis was in some measure the fault of the Americans." Most likely Khrushchev "would trade Berlin for Cuba any day," which would be a "heavy net loss" for the United States. Under a prolonged blockade, the morale of Berlin would likely crack, and no one would fight a nuclear war for a "dying city." Therefore, Klein advocated that the United States "shorten our reaction times" in the Poodle Blanket plan, "and be prepared to confront Khrushchev at a very early stage with a bluntly nuclear choice."

For Dean Rusk, the tension of indecision was stifling. Taking the elevator from the State Department parking garage to his seventh-floor office with George Ball and former secretary of state Dean Acheson, Rusk pointed to his security guards and said, "You know, the only decent advice I have had this past week has come from these two fellows." One of them, a former pro football lineman, responded, "The reason for that, Mr. Secretary, is that you have surrounded yourself with nothing but dumb fucks." Rusk laughed, as Ball and Acheson blushed.

LESTER Bernstein had not returned to Germany since his army tour at the end of World War II. As a Jew whose immigrant parents still spoke Yiddish, his emotions had to be complex as he landed in Berlin. His mission as NBC's fixer, at least, was clear: to save the tunnel film. But the trip got off to a rocky start when he could not arrange a meeting with Willy Brandt. Bernstein did manage to outline for Egon Bahr, the Brandt aide, and leaders of the Senate the propaganda value of the film and the steps the network was (reluctantly) taking to hide the identities of certain tunnelers and escapees. Still, the Germans remained skeptical. They even added a new argument: the East Germans in response to the film might increase the prison sentences meted out to escapees and couriers involved in *other* tunnels.

Bernstein had hoped to spend just a couple of days in Berlin, but that was out of the question now. On October 19 he wrote to his mother apologizing for not telling her about this "special assignment" that had suddenly taken him to Europe, and which "seems to be stretching out somewhat." After briefly describing the tunnel film, he revealed, "The problem now is that the West Berlin government has asked NBC not to show the program," so he was trying to persuade it "to withdraw its objection. . . . I'm hopeful that I will succeed in this."

The same day, Bernstein sent a memo to Egon Bahr arguing

that NBC would seriously consider any "clear and specific risk"
their program might pose. Cannily, he added:

> Frankly, the NBC management also finds it difficult to take
> very seriously the government's concern over publicity on
> tunnel digging at a time when the government is cooper-
> ating wholeheartedly with a gala premiere of the movie
> *Tunnel 28* almost in the shadow of the Berlin Wall itself.
> Moreover, we feel it is unfair for the government to single
> out and discourage the NBC project—which is a responsible,
> authentic presentation of actuality—after going out of its
> way to make possible the filming of a fictional, romanticized
> treatment of the same subject.

And Bernstein didn't even know about the West Germans helping a fake tunnel escapee fly to America to promote the MGM
film.

Bernstein concluded, "NBC wants to emphasize its convic-
tion that the showing of its program will be very much in the
interest of West Berlin and all others who oppose Communism."
Called out for its MGM vs. NBC hypocrisy, the German opposition faded quickly. Officials told Bernstein privately that they
would no longer fight the film, although they offered no specific reason for the reversal. Senate leaders issued a statement
that they had received from NBC assurances that the film "does
not represent any danger to the safety of the participants." They
added that they respected the fact that "NBC is guided by the
wish to bring an authentic report to the public of the United
States in the interest of Berlin."

When Lemmer, the federal minister for all-German
affairs—the same official who had personally arranged the
Tunnel 28 premiere—said on October 22 that he would welcome
the NBC film "if the events in Berlin were reported to the world
public as extensively and precisely as possible," Bernstein cabled
the news to his home office, then made sure the wire services
received the statement. In short order UPI declared the German
objections "ended."

Would this sway the U.S. State Department? It no longer mattered, at least in the short run. NBC's Kintner had just postponed the airing of *The Tunnel* indefinitely. Reuven Frank feared it was worse than that—the program could be dead. After a chat with Bill McAndrew, he suspected that someone in an even higher corporate position had gotten to Kintner with a warning that he was not to put in jeopardy the many lucrative U.S. military contracts enjoyed by NBC's parent, the RCA corporation. Frank felt so depressed he pondered whether he should stay at the network whose news division he helped build.

TWO major articles about Berlin after the Wall now arrived in U.S. magazines. A sprawling piece for *The New Yorker* by John Bainbridge, "Die Mauer: The Early Days of the Berlin Wall," covered all the bases, from the day the barrier emerged to escape attempts. The longest take on the subject ever to appear in America, it featured visits to various locales along the Wall, plus the refugee center at Marienfelde. Bainbridge reported that about one in twelve escapees in the past year were former border guards or East German soldiers. With "scarcely an exception" the deserters listed as a primary reason the "order requiring them to shoot at would-be refugees." Recognizing this, the GDR had reassigned most of the Berlin-born guards and replaced them with soldiers from the hinterlands.

Renowned foreign correspondent Flora Lewis, meanwhile, produced a feature for *The New York Times Sunday Magazine* comprised of mini-profiles of Germans who professed so much love for West Berlin, despite its isolation and dangers, that they would never leave. "It is not so much that people have changed under the increasing pressures," she explained. "Rather, they have been stripped and things show more—fears, ideals, jealousies, heart." She could have been referring to the tunnelers and escape artists.

One of those interviewed was James O'Donnell, the writer and former aide to General Clay who had introduced Daniel Schorr to the Kiefholz tunnel. Why did O'Donnell return to the city again and again? "Because I'm an American and I believe in it," he replied, "and if we won't do what we believe in, we'll go down."

Another profiled was "Horst P.," a twenty-one-year-old former theology student turned "sandhog" on the new "underground railway" run by modern-day "Scarlet Pimpernels." He had left East Germany, alone, at age fourteen. Lewis found him waxing indignant the night she interviewed him in a busy pub. His tunnel crew had planned to rescue a group of people about to be arrested in the East, and had figured that the landlord of a building near the border would surely let them use his basement, especially when they offered money. "But all he could think of was his leases and his contracts," Horst P complained. "He found a million excuses. His tenants might be inconvenienced. The Communists might shoot across the Wall. Though he never said it, I think underneath he wanted to be sure that in case the Russians take over one day he would have no mark of resistance on his record. Just like the Nazi times—all these good brave bürgers wanted was to say, if the Gestapo came, 'I wasn't involved.' . . . They are not traitors, these landlords, but they are cowards, and that's no better.

"I think that anybody who isn't ready to help someone in trouble for lack of freedom doesn't deserve his own freedom. And I mean not just talk about it, do something about it."

THIS autumn, nuclear fear was in the air almost everywhere in American culture. The latest edition of *Esquire* magazine featured on its cover an illustration of a modern Noah's Ark—a basement fallout shelter. Crammed inside the cinder-block space were a nicely dressed man and woman, plus two lions, two chickens, and two llamas. A young folksinger named Bob Dylan was starting to perform live a song he had written a few months back warning of nuclear war. Its title: "A Hard Rain's A-Gonna Fall."

In his book *Run, Dig or Stay*, Dean Brelis confronted the shelter mania sweeping the country—mostly in the media, as actual construction lagged—before concluding that the lofty hopes and ideals of America "are no good down there in the ground." Seymour Melman, a Columbia University professor, edited *No Place to Hide*, a popular collection of pieces raising doubts about civil defense. Meanwhile, the Pentagon released figures, probably exaggerated, claiming that "potential" fallout shelters existed in some 112,000 structures across the land, with room for more than 60 million Americans. At a meeting with newsmen, Robert McNamara discussed how civil defense offered protection in a missile attack. "The newsmen roared with laughter and derision," the *New York Times* observed.

Nuclear terror might have been sparking more fatalism than activism, but a new novel seemed poised to effect both. *Fail-Safe*, written by Eugene Burdick and Harvey Wheeler, portrayed an accidental (but inevitable) nuclear attack sparked by U.S. bombers mistakenly receiving a "go" to attack the Soviet Union. It was news to most Americans that many of their Strategic Air Command bombers with nuclear payloads were airborne at all times, ready and willing to devastate the Russians at a moment's notice. *Fail-Safe* was an immediate sensation, the subject of serious news stories and commentary. (The book bore a strong resemblance to a 1958 novel with a similar plot, *Red Alert*, by Peter George, who was now working on a movie adaptation with director Stanley Kubrick.) *Fail-Safe* ended with the United States still standing—the Soviet leader choosing not to retaliate after America's accidental first-strike attack. Why? Only because the U.S. president, to show contrition and good faith, ordered his military to drop a single nuclear bomb on Manhattan, where his wife was visiting.

No wonder U.S. officials and military leaders were afraid the book would increase public paranoia, leading to demands for cutbacks in their nuclear arsenal or an end to hair-trigger alerts. A reviewer for the *Chicago Daily News* seemed to have this in mind when he declared that the book "can vitiate our will and warp our judgment. It must not be allowed to do so." The Air

Force had ordered a fact-checking study of the novel and an estimate of how much more damage it could do if, as expected, it became a major Hollywood movie.

Unnamed "administration officials" tried to debunk the *Fail-Safe* scenario. The authors had "distorted" the picture, they charged, making alerts seem far more hair-trigger than in reality. The same officials volunteered, however, that the military was now spending vast sums of money to take further precautions against accidental war. The Pentagon, a *New York Times* article concluded, was hardly "complacent" about the threat because—and here was the kicker—they did not consider current precautions close to "foolproof."

WITH his corporate diplomacy completed, NBC's Lester Bernstein on October 22 attended one of the events of the year in Berlin, the premiere of MGM's *Tunnel 28* at the swank new Kongresshall in the Tiergarten. *Variety* noted that this was the third world premiere of an American movie in the past twelve months in Berlin, following Billy Wilder's *One, Two, Three* and Stanley Kramer's *Judgment at Nuremberg*. Wilder's movie bombed at the box office but it did include one prescient scene, with a Russian official handing James Cagney a Cuban cigar. "We have a trade agreement with Cuba," the Russian explained, "they send us cigars, we send them rockets."

And what a gathering at the Kongresshall! Also in attendance, besides MGM execs, *Tunnel 28* creators and stars, and government officials, was Rainer Hildebrandt, who had just opened the first exhibition of Wall-related memorabilia in Berlin. He brought as his guest none other than one of the world's leading experts on escape tunneling: Harry Seidel. Heinrich Albertz, minister for internal affairs, refused to attend, objecting to what he felt was the exploitative nature of the MGM movie. He was rebuked by fellow minister Ernst Lemmer, who hugged producer Walter Wood, saying, "I thank you in the name of the German people for this statement against inhumanity."

The Berlin Hilton hosted a gala afterward, complete with festive dance music. Bernstein was chatting with NBC cameraman Harry Thoess when Daniel Schorr came by for a little chat. Schorr immediately started needling them about "checkbook journalism" and that week's State Department attacks on the NBC film, in contrast with his own network's decision to toe the line.

Schorr, of course, wasn't exactly on firm ground. He had bitterly *opposed* CBS's playing ball with Dean Rusk on the matter—and he, too, had paid tunnel organizers for the right to film their operation. The diminutive Thoess would have none of it. He denied the State Department claims about the Bernauer tunnel and told Schorr he knew all about his scheme to film the Kiefholz escape. Sputtering, Schorr denied any involvement, but Thoess said NBC had proof, in the form of the footage shot on August 7 by Piers Anderton. He claimed it showed a CBS cameraman trying to film across the border fence that day— presumably the man Anderton had noticed crawling through bushes. What would CBS and the State Department say if NBC released *that* footage? At this point Bernstein introduced himself as the network's legal fixer from New York. The NBC men might have only been bluffing, but Schorr turned and walked away at once.

AS the film party in Berlin ended, President Kennedy, four thousand miles away, was getting ready to deliver a nationwide television address. In discussions and debates with top aides and military advisers over the past two days, Kennedy had stuck to his decision to issue an ultimatum to the Soviets to remove all of their missiles in Cuba immediately, along with the imposition of a blockade of the island, before taking military action. But he would wisely refer to the blockade as a "quarantine," to distinguish it from the only blockade most Americans were familiar with—Berlin, 1948, by the Soviets—while communicating that only offensive weapons, not fuel, food or medicine, would be halted. Forty U.S. ships were already in position to enforce it. In any case, if the Soviets challenged the blockade or flatly ruled out removing the missiles, U.S. air strikes would arrive swift and strong.

A few days had now passed since the *Washington Post* learned about the discovery of nuclear missiles in Cuba, but the President had asked editors to hold off on revealing this until his televised speech, and the *Post* complied. Other media fanned the fears provoked by rumors. Lester Bernstein's family in New York, knowing that he was still in Berlin—sure to be one of the world's most endangered cities in coming days—grew particularly worried. And they didn't know that Berlin Mission chief Lightner and at least one other official there had already evacuated their wives and children to America.

As the afternoon wore on, U.N. ambassador Adlai Stevenson cabled Dean Rusk: "I believe decisions must be made before this evening about how we will react if the USSR in the next few days seeks to interfere with our access to Berlin. Decisions need to be made both with respect to our physical reaction in the area and what to do in the UN." President Kennedy was also concerned about the chain of command if the United States was tempted to respond to a Soviet attack by firing nuclear missiles. He assumed that the American missiles closest to Russia, in Turkey and Italy, could be fired only after a specific order from him. This turned out to be untrue. The local U.S. commanders had been given that power under standing orders. When JFK told one of the Joint Chiefs that afternoon that he didn't want *any* missiles fired without his express approval, the general seemed to balk. A few minutes later, when the President brought this up with Paul Nitze, an assistant secretary of defense, he again met resistance.

Over the course of the day it became clear that Kennedy and top administration officials were not completely up to speed concerning American nuclear forces and their possible use. Dean Rusk, for example, asked Nitze, "The ones in Turkey are not operational, are they?"

"Yes, they are," Nitze said.

"Oh, they are," Rusk said, dryly.

"Fifteen of them are on alert right now," Roswell Gilpatric added.

Seconds later, Nitze pointed out that NATO policy called for the automatic launch of such missiles in the event of a Soviet strike, "according to its EDP."

"What's EDP?" President Kennedy inquired.

"The European Defense Plan," Nitze responded, "which is nuclear war."

Kennedy coolly replied that this might be fine if it came to a wide-scale Soviet attack, but not a "spot" one. He insisted again: *Tell the commanders that I am in charge.* This time Nitze said he'd do it.

In a meeting to inform congressional leadership about his coming speech, JFK listened to Richard Russell, the top Democrat on the Senate Armed Services Committee, argue for a U.S. strike on Cuba, asserting that the nation had to "take this gamble" even if it risked nuclear war. The leading Democratic senator on foreign policy, J. William Fulbright, slammed the blockade and called for "an invasion and an all-out one."

At 7:00 p.m., the President went on the air with his historic "quarantine" speech, as written, mainly by Ted Sorensen. He referred to Berlin only briefly but with emphasis: "Any hostile move [by the Soviets] anywhere in the world against the safety and freedom of peoples to whom we are committed—including in particular the brave people of West Berlin—will be met by whatever action is needed." And then he, and the world, waited.

WHEN he awoke the next morning in his sunlit bedroom, Dean Rusk thought, *Ah, I am still here.* The nation had survived, so far.

NBC used the occasion of the Cuba crisis to make official something executives had already decided. "In view of the critical international situation that has developed in the last twenty-four hours," NBC announced, it had concluded "that this is not

an appropriate time to broadcast its documentary program showing the construction of a tunnel under the Berlin wall and the
escape of East German refugees." *The Tunnel* was postponed, if
not canceled.

When he received the news directly from Robert Kintner,
Reuven Frank sat silent for a long time. One might argue that
this was exactly the time to air such a report. With an American
swap of Cuba for Berlin likely to be on the table very soon, if it
wasn't there already, the future of a great city (or at least half
of it) was up for grabs. Outraged, Frank told Kintner that he
was just using the Cuba crisis as an excuse and that he sensed
NBC had planned to kill, not just postpone, the program ever
since it became politically controversial. Frank demanded that
he at least be assigned to work on the ninety-minute documentary now set for "my time slot"—a special on the missile crisis.
Kintner agreed. This was a program not likely to be killed due
to White House or State Department pressure.

In West Berlin, newspaper editorials and officials hailed
Kennedy's speech. Ordinary Berliners seemed to feel the same,
with many encouraged that the United States "has finally taken
vigorous and positive action against Communists," as one article
reported. A top CIA official at the Office of National Estimates
sent a memo to his director asserting that West Berlin could
hold out, physically and economically, for quite a while. It had
stockpiled enough food and medicine for at least six months, as
well as enough fuel for a year.

On October 27, a Soviet submarine off Cuba, cut off from
communicating with Moscow, nearly fired a nuclear-tipped
torpedo after mistaking depth charges dropped by a U.S. ship,
meant to force the sub to the surface, for the start of an American attack. Two officers approved the firing of the torpedo. A
third did not, and after a heated argument he prevailed, possibly
preventing a nuclear holocaust and World War III.

Later that day, after nearly two weeks of tension and despair,
warnings and threats, a deal to end the missile crisis was finally
in place. Its terms had been fairly self-evident from the start.
President Kennedy agreed to remove the obsolescent U.S. mis-

siles in Turkey in exchange for the Soviets dismantling all missiles in Cuba—but only if Washington's part of the deal was not announced publicly; the U.S. victory must be portrayed in the media as complete. The nuclear retreat from Turkey, JFK vowed to the Soviets, would quietly happen a few months down the road. The United States also promised not to invade Cuba, nor to assist exiles bent on trying that again.

The following day, around nine o'clock in Washington, Nikita Khrushchev broadcast a message on Radio Moscow. His government, he said, had issued an order to start dismantling the missiles. General LeMay nevertheless told President Kennedy that he had "been had" and called the settlement "the greatest defeat in our history"—and continued to urge a U.S. invasion of Cuba.

American spy planes soon discovered, to the surprise of many, that the Soviet dismantling was indeed taking place. The crisis was over. West Berlin, after all the fear and expectation, was neither traded nor seized. But the Wall remained, and Khrushchev's promises to escalate conflict there remained in place. On October 30, however, the *New York Times* reported that senior Western officials were claiming that Khrushchev, in the wake of his humbling setback on missiles in Cuba, would now "proceed cautiously" in Berlin. That's because the Western allies had revealed "unity" and the United States "resolve" in meeting the threat, which signaled they would do the same in any Berlin crisis.

President Kennedy wrote *New York Times* publisher Orvil Dryfoos, who was still smarting from the Hanson Baldwin affair, to thank him for ordering his editors to hold off reporting on Soviet missiles in Cuba before his October 22 speech. "Events since then," Kennedy declared, "have reinforced my view that an important service to the national interest was performed by your agreement to withhold information that was available to you."

UNLIKE the NBC film, MGM's *Tunnel 28* had reached the screen right on time, at least in Berlin, but the early reviews were brutal. *Der Spiegel* compared it unfavorably to the real-life newsreel footage that opened the film. *Die Zeit* charged that while it had "artistic value," as a political document "it is a scandal." Apart from a young mother who wanted her baby to enjoy a life of greater opportunity, "there really is no one among the refugees who has a reason to flee." It quoted one of the female characters dreaming of a chic Western haircut and more: "I want to go where I can take a hot bath" with "thick terry towels to dry me." So now, the writer observed, American mothers will know why their sons might fight and die for Berlin: "for the hot bath and haircut."

Director Robert Siodmak defended the film's pedestrian approach to history and politics, telling *Die Zeit* that it was made for illiterates who didn't know what the Wall was. The newspaper commented sarcastically: "Now they know it so!"

The plot was highly fictionalized from the Becker saga. Kurt (Don Murray) works as chauffeur for a Major Eckhardt, while having an affair with the major's wife. Then Kurt's friend Günther dies while attempting to crash through the Wall in a truck. Günther's sister Erika (Christine Kaufmann) believes he made it to the West and tries to join him. Kurt saves her from getting shot and whisks her into hiding at his apartment, right at the border. He is not interested in exiting himself, but offers to help dig a tunnel, starting under their building, so that others can leave. Kurt falls in love with Erika and finally tells her that Günther is dead. Naturally he now wants to join her in escaping. Erika's father, a Communist, finds out about her plans and tells Major Eckhardt about the tunnel. Kurt urges the swelling number of refugees to flee immediately! Police and VoPos arrive a little too late but shoot Kurt, the last one in the tunnel. He manages to crawl to freedom, where his reward may be a happy future with Erika.

The movie was set to open in the United States on October 31, under a different title: *Escape from East Berlin*. In Hollywood, MGM sent out publicity releases linking the film not to the Becker escape (which it had bastardized) but to the more recent Bernauer tunnel. "Not long after *Escape from East Berlin* was completed," one release boasted, "newspapers all over the world announced the news of another successful escape through a tunnel under the Berlin Wall, this time achieved by a group of 29 East Berliners." MGM updated its bright-red film posters with the new title and actual headlines from coverage of the Bernauer escape, such as "Reveal 29 Fled Reds in Tunnel." In large letters: *It's Happening As You Read This.* The campaign gave literal meaning to the familiar claim that a movie was "ripped from the headlines."

In the aftermath of the Cuba crisis, however, MGM decided to open the film only in Michigan (at first). Perhaps the studio, observing what had happened with the CBS and NBC documentaries, feared pressure from the State Department. Producer Walter Wood, for his part, told a newspaper that he didn't believe in making "exploitation films." If he had wanted to do that with his Berlin tunnel film, he said, he would have shown twenty-eight deaths "rather than twenty-eight escapes."

17
Sabotage

NOVEMBER 1962

More than a year had passed since Harry Seidel announced to friends and family that he would rescue his mother from the East, or die trying. So far, with his mother behind bars most of the time, his exertions had come to naught. Now he learned that his mother had finally been released from prison. Fritz Wagner was having trouble getting another operation going, so when Harry heard about a tunnel already in progress in the Kleinmachnow neighborhood far to the west in the American sector, he volunteered to finish it.

Separately or together, the organizers of this new dig, the brothers Franzke—Boris and Eduard—had been digging tunnels at the border for over a year. They wanted to retrieve from the East their sister and mother; Boris's fiancée; and Eduard's wife and two children. In the autumn of 1961 Eduard had attempted, and quickly abandoned (after attracting Stasi attention), perhaps the first tunnel project in Berlin after the Wall went up. Boris then helped him escape to the West via a window at an S-Bahn station. Not one of the several tunnels they had started since had come to fruition. That hardly discouraged them.

In October 1962 they heard about a family named Schaller just over the border who wanted to flee, and who were prepared to offer their home in the leafy outskirts of Berlin as a tunnel destination. They jumped at the opportunity, as usual. The man who tipped them off, Bodo Posorski, operated a car-rental business near the Franzkes' home and offered to help them with the new tunnel, as he had with other escape operations.

In the Kleinmachnow area there was no imposing wall, just a series of fences and barbed wire, with relatively few patrols and modest surveillance. The Schallers (father, mother, two daughters) lived on Wolfswerder, the first block from the border. Across the fencing in the West a residential construction site buzzed with activity. This might confuse or distract border guards in the East. Finding no structure in the West whose basement could be borrowed, the Franzkes had this bright idea: rent a pair of good-size sheds, truck them to the construction site, and place them on the edge of the property, just yards from the border. One could be used for supplies, the other for sleeping. Then post a sign declaring this was the headquarters for a new landscaping business: *Evergreen Gardening*.

This ploy would account for people (diggers) coming and going, and the delivery of equipment such as shovels, wheelbarrows, and axes. Even so, the Franzkes asked two colleagues, Bibi Zobel and Klaus Gehrman, to join them in bunking down in one of the sheds, sleeping on inflatable mattresses, for the three weeks needed to tunnel to the Schaller residence. Bodo Posorski periodically dropped off food and water. It was no easy task speaking only in whispers, using flashlights for illumination, and enduring the confined smell of unwashed bodies, but the work went well. They even connected a phone line in the tunnel, courtesy of a modest advance from *Bild Zeitung* in exchange for (what else?) their story rights. The supply shed also held the sacks of dirt removed from the underground.

By the end of October, however, the declining levels of air in the chamber slowed progress. Diggers complained of headaches and could work only in short bursts, growing shorter each day. The tunnel had to be finished soon. Too many people already knew about it: the Schallers had invited friends to join the escape and couriers managed by Posorski had spread the word to about three dozen others (some of whom offered payment). The Franzkes decided they needed an experienced digger to expedite the project.

A suggestion from Posorski led them to the peerless, and fearless, Harry Seidel.

The Franzkes were considered a rival blue-collar tunnel team by Seidel's usual boss, Fritz Wagner, but Harry was ecumenical. Also, his mother was now free to flee. When he arrived at the tunnel site on the afternoon of November 5 and took a look around, he said little beyond expressing pleasure that they had advanced so far already. He went to work immediately, showing his usual prowess in low-oxygen conditions. The Franzkes offered to let him go home to his wife when he finished a shift, but Harry, figuring this was a one-week project, chose to stay in the makeshift barracks with the others. When Harry told Wagner what he was doing, *Dicke* said that this wasn't news to him— and that too many others knew about it also. He advised Harry to quit before it was too late. Harry said, "I'll think about it," which in Seidel-speak meant *no*. He had promised his son that the boy would see his "granny" before Christmas.

The Franzkes estimated that the tunnel needed to extend 250 feet to hit the Schaller house, so when they reached about 230 feet they figured they were at the outside wall of the house. So they dug a small hole into the open air and took a quick look around. *Damn, we're still in the middle of the front yard.* They would have to dig on. Unfortunately, Bodo Posorski had already told couriers that the escape action would start on November 11, with other groups arriving in the days that followed. The Franzkes decided to shovel just a few more feet and, in a day or two, break through just outside the house, after it grew dark. One of them would squeeze through the hole, look for a prearranged all-safe signal in the front window of the house, knock on the side door and invite the waiting refugees to follow him to freedom. Risky, but maybe safe enough in this quiet, remote area.

What they didn't know was that the Stasi had been tipped off to the tunnel and the escape plans. Agents had already arrested the Schallers and their two daughters, plus nine early arrivals, on November 11. Now they were monitoring the path of the tunnel with listening devices, with plans to blow it (and maybe some tunnelers) to kingdom come at an opportune moment. A Stasi explosives expert was at the scene. After months of frustration, they finally had the infamous Harry Seidel and the bothersome Franzkes at their mercy.

HOPING for more arrests, the Stasi had set up a virtual armed camp at the Schaller residence on Wolfswerder. Under pressure, the Schallers had told the Stasi about the "all-clear" signals they had been sending to the Franzkes—cleaning windows with a dust cloth every morning at ten o'clock and chopping wood outside at 4:00 p.m. The Stasi continued the charade. One agent even dressed as a woman for the window cleaning.

Lieutenant Richard Schmeing, a fifty-three-year-old MfS explosives chief, took charge of more lethal preparations. Blowing up the tunnel at Wolfswerder, especially with Harry Seidel on the scene, was such a high priority that Stasi director Erich Mielke had personally approved the plan. Even if the MfS didn't exterminate Seidel and the Franzkes, it could destroy the chamber and then claim that the "terrorist" tunnelers had set off the charge in a barbaric attempt to destroy lives and property in the East.

The day after the first arrests, Schmeing's team, out of view from the West, dug a hole between two houses across the street from the Schaller house, right at the border, directly over what they surmised was the path of the tunnel. They inserted two packs of explosives—2.5 kilograms of TNT, and the same amount of RDX—and covered them with autumn leaves. The explosives were powerful enough to destroy many yards of property, so the location didn't need to be exact. When there was no escape action that night, November 12, Schmeing removed the charge. This was repeated the next day. One of the couriers, growing nervous about the operation and worrying that Bodo Posorski might not be on the level, asked her mother to visit the Schaller house. When she arrived, the stranger who answered the door said the Schallers were away (he did not reveal that they were "away" in prison). The woman left and quickly told the courier that the Stasi may have raided the house already. This was allegedly passed along to Posorski, who either dismissed it as fantasy—if true, why hadn't the Stasi arrested the woman?—or ignored it for some other reason.

On November 14, Schmeing determined via listening devices that the diggers were moving equipment to the front of the tunnel, indicating an impending breakthrough. He adjusted the placement of the explosives and ran copper wire two hundred feet into the basement of the Schaller house, linking it to a 12-volt dry battery attached to the detonator he would likely, that night, finally push to destroy the tunnel and everyone in it.

By 8:00 p.m. darkness had fallen. The temperature was barely above freezing. A few more clueless refugees had just been seized and either hauled away or kept under guard in the Schaller house. (Harry's mother and Boris Franzke's fiancée had, fortunately, not yet arrived.) Stasi agents were stationed around the property. Lieutenant Schmeing stood at the basement window facing West, watching and waiting, as his Stasi comrades flashed one of the bogus all-safe signals to the tunnelers.

Inside the tunnel, just yards away, the diggers debated who would actually climb through the hole into the front yard. The Franzkes argued, and to Seidel it seemed neither particularly wanted the assignment. As was his wont, Harry took charge. He had promised his friends and his wife that he would never be first out of a tunnel again, but now declared, "I'll do it." Harry wrapped his pistol in plastic to protect it from the sandy soil, climbed on the shoulders of big, strong Bibi Zobel, and was out the hole.

After Harry exited, Boris Franzke mounted Bibi's shoulders and stuck his head just above the earth, keeping watch, with his old Wehrmacht gun drawn in case Harry started taking fire. He saw Seidel approach the house and disappear behind it. Out of sight, Harry climbed the steps to a terrace and rapped on the door with his weapon. When someone opened the door, Harry saw not an anxious escapee but a squad of heavily armed Stasi in civilian clothes and soldiers in uniforms. Faced with machine guns, Harry dropped the pistol, as he was shoved to the floor, kicked, and pummeled. Stasi agents frog-marched him outside, weapons aimed at the jagged hole in the ground.

The Franzkes, just inside the tunnel, heard a familiar voice exclaim, "Come, we need to help a sick person!" Boris was about to climb out of the hole but his brother held him back. The tunnel team had conversed in nothing but whispers all night. Now here was Harry speaking in a suspiciously loud voice, totally inappropriate with East German guards stationed nearby. They heard him repeat the request, again in that odd tone. Was Harry trying to warn them?

Seidel, hearing a stir inside the tunnel, suddenly shouted, "Go away! The tunnel is betrayed, soldiers will shoot you in the head!" The Franzkes took the hint, and began a mad scramble to the West. Harry was knocked on the skull with a pistol and roughly hauled to the Schaller house, where he suffered another beating.

Inside the laundry room in the basement, it was time to blow up the tunnel. The Stasi commander on the scene, Lieutenant Colonel Siegfried Leibholz, ordered Lieutenant Schmeing, *"Sprengen!"* ("Ignite!") There was just one problem: two teenagers from the neighborhood were talking and maybe kissing in the dark no more than a dozen yards from where the powerful explosives had been planted. They had returned from seeing a movie, and seemed unaware of the noisy activity across the street. It didn't look like they were in any hurry to go home.

"Look! The young lovers!" Schmeing protested.

Leibholz was insistent: "Ignite!"

Schmeing, his back to Leibholz, hesitated. The explosives expert had survived two Nazi death camps, and had been slated to be part of a deadly typhus experiment at the Buchenwald camp when the war ended. He may have had more moral qualms than many knew. Finally he pressed the detonator.

Nothing happened. He tried again—same result.

Near the tunnel's exit hole, Bibi heard the frantic cry of "Ignite!" and a moment later, from the same direction: "The pigs are escaping." With that, he hastened to join the panicked crawl back to the West.

Minutes later, outside their two sheds in West Berlin, the agitated Franzkes told their colleagues what had happened as they lit torches. There was still no sign of Harry Seidel. Bodo Posorski arrived and asked the Franzkes for one of their guns to fire across the border at the VoPos. Nearby residents noticed the unusual commotion and called the West Berlin police. As was customary, officers investigated but did not arrest anyone or charge them with weapons possession. They did seize the bags of soil the diggers had piled in the hut—to use for police target practice.

Harry Seidel, in custody across the way, was wondering if Bodo Posorski had organized this tunnel, and insisted that he join the Franzke crew, under the influence of the Stasi. In fact, a young man named Werner Kiontke, who sometimes managed Posorski's business when he was away, happened to be a newly minted Stasi informer. In early September, Kiontke, who had a West German passport, had been arrested for a traffic offense in the East. The Stasi had induced him to reveal everything he knew about the "criminal" activities centered on the car rental office, and to monitor it further, as a secret informer with the code name "Roge." He was clearly in position to betray the Franzke tunnel by spying on or manipulating Posorski.

BACK in the East, Schmeing, defying an order from Leibholz to remain in the house, went out to investigate why his bomb charge had fizzled. He picked the copper wire off the ground and followed its path across the yard—until he came to where, he later reported, it had been crudely cut, perhaps with a knife or piece of glass, and pulled apart. Someone on the Stasi team had apparently engaged in an act of humanitarian sabotage, perhaps a Stasi first. There would be hell to pay, especially since the much-feared MfS director Mielke had himself taken such an interest in the detonation.

Returning to the house, one of Schmeing's comrades told him to not feel so bad. Yes, he might suffer because of this snafu. On the other hand, there was also the chance that he had saved himself from a long prison term. Some in the East might not be as happy as Mielke about writing off two innocent teens as collateral damage.

Seidel, meanwhile, was hustled to the Stasi prison at Hohenschönhausen, where he was interrogated all night. The MfS hastily drafted a five-page memo describing all of his known tunnel projects in the past year. (They falsely claimed thirteen in all.) The following morning one of the diggers rushed to Harry's apartment to tell his wife what had happened. Later the West Berlin police dropped off a bundle: Harry's clothes, which had been left in the shed. Two days after that, Werner Kiontke gave a Stasi agent the key to Posorski's car-rental shop so they could make a copy, enabling them to return later and plant a listening device.

PRESIDENT Kennedy was feeling more politically secure than he had in some time. He had drawn wide praise for his handling of the Soviet missiles in Cuba, and this seemed to have helped some Democratic candidates in the midterm elections. As usual in these races, the party holding the White House lost a few seats in the House, but the Democrats maintained their majority there while widening their edge in the Senate. Another of the President's brothers, Ted, had been elected to JFK's old Senate seat in Massachusetts.

In solidifying his triumph in Cuba, however, the President had sparked wide resentment among the media. Many felt that the White House had manipulated, even lied to, the press while the crisis was happening. (And they still didn't know about the secret agreement to bring U.S. missiles home from Turkey.) Others resented Pierre Salinger's repeated requests for self-censorship during the crisis, along with his twelve-point list of "guidelines" for withholding news. Surely now, with the crisis over, the administration would admit it went a little too far, or at least shed its crisis-born demands for secrecy.

Instead, Pentagon spokesman Arthur Sylvester set off a firestorm when, pressed by a reporter, he admitted that the administration's control of information was even tighter than during World War II—yet defended it, due to "the kind of world we live in." It was important for the nation to speak with "one voice to your adversary." He used a loaded new term in speaking favorably of government "management" of the news. Journalists of all political persuasions raised a hue and cry, declaring that they were now expected to act as little more than government propagandists.

The *New York Times* declared in an editorial that "management" or "control" of the news "is censorship described by a sweeter term." The *Times*'s legendary Arthur Krock, while admitting that all presidents tried to manage the news, opined that "direct and deliberate action has been enforced more cynically and boldly" by this White House "than by any previous administration when the U.S. was not at war." The *Washington Star* asserted that Sylvester had "let the ugly cat out of the bag," calling his comments, "truly sinister." George Sokolsky, a conservative newspaper columnist, compared current U.S. press control with that under Hitler, Stalin, and Castro. *Newsweek* was more generous, deciding that if the White House did indeed "mislead the public, it accomplished its tactical end—allowing U.S. strategists to work in secrecy. . . . It was part of the grand strategy by which the U.S. had risked nuclear war—and won without one."

Sylvester's views were largely shared at the White House. Kennedy himself had used the phrase "news management," and Salinger believed that disinformation and even lies were justifiable measures in a conflict with an enemy who had the advantage of operating in secret. In a clarifying statement, Sylvester said he respected the journalists' concerns, but emphasized that national security and the safety of U.S. personnel required careful scrutiny of the media. Sadly for Sylvester, the *New York Times* chose to characterize his remarks with the headline, "U.S. Aide Defends Lying to Nation."

Privately, JFK admitted to his friend Ben Bradlee, who was now editing *Newsweek*, that the U.S. had indeed "lied" to the press during the Cuban missile crisis. And the policies the President had ordered in the wake of the Hanson Baldwin episode to monitor both reporters and government sources had just been instituted, at least in part, at the Pentagon and in the State Department. McGeorge Bundy wrote to columnist Joseph Alsop, "We are aiming at dangerous reporting assisted by irresponsible or careless officials.... [T]his kind of reporting exists, and ... there are such officials."

NO word had emerged on the ultimate fate of *The Tunnel*, beyond a brief mention in *Variety* that the program would "probably" be aired someday. The same story credited NBC's Lester Bernstein with helping improve German officials' attitude toward the film during his recent visit. It disclosed nothing on the current views of their American counterparts, however. Beating NBC to the punch with a tunnel program, and achieving some measure of revenge, CBS's *Armstrong Circle Theater* aired a one-hour docudrama titled *Tunnel to Freedom*. It fictionalized the so-called Pensioners Tunnel of the past May, when elderly Berliners had dug an unusually spacious shaft so they could walk erect, toting suitcases, into the West. *Variety* judged the reenactment fairly accurate if not exactly "exciting." Conrad Nagel played the lead.

Reuven Frank had waited long enough. After much consideration and consultation with his wife, he decided to write out and submit his resignation. He believed that NBC had held the fort against the Kennedy administration longer than competitors

might have, but he still resented *The Tunnel* postponement—
which he felt sure would be permanent. He questioned whether
he could stay in an industry where "political" (as he called it)
pressure would kill such a worthy endeavor. Bill McAndrew
asked him to at least put off his departure until the end of the
year so he could complete two programs in progress.

Piers Anderton, similarly fed up with his network's *Tunnel*
collapse, finally left with his wife for a monthlong honeymoon
in Greece and Italy, delayed since June.

On November 20, Robert Kintner, the NBC president, sent
Secretary of State Rusk (carbon copy to Pierre Salinger) an
unprecedented five-page, single-spaced letter, along with an
eight-page memo, to "set the facts before you in some detail"
concerning the escape film that had "been the subject of so much
confusion, misinformation and misunderstanding." Kintner defended the program partly by providing false information, such
as "completion of the tunnel was in no way contingent upon the
payments we made." There was no mention of the NBC apartment's crucial role in signaling couriers and refugees on escape
night. He pointed out that the State Department never explicitly
asked NBC to halt the project, which was true, but only because
State did not know about it.

Kintner warned that NBC still hoped to schedule the show:
"In our view, having successfully overcome whatever risks our
undertaking entailed, it would be folly indeed to deprive the
cause of freedom—as the Communists would wish—of the
clear advantages to be gained by televising the film." *The Tunnel* would give "millions of Americans a keener awareness of
their stake in Berlin and a deeper insight into the nature of the
struggle between Communism and freedom." He quoted Secretary Rusk's own speech during his June visit to Berlin when he
said of the Wall: "It is an affront to human dignity."

Unable to refrain from exposing a rival, the memo also informed Rusk that contrary to its public statements, CBS had not
withdrawn from filming at Kiefholz Strasse at State's request.
NBC had "reliable information" that a CBS cameraman was
"stationed at the West Berlin mouth of the tunnel" on escape day.

NBC's Bernstein and Elie Abel personally delivered the Kintner letter and memo to Rusk. After receiving his copy at the White House, Pierre Salinger wrote to Kintner, "I understand you will be hearing directly from the Secretary but I wanted you to know I was grateful to have your views." There was this important twist: Kintner told the State Department's Robert Manning that he hoped to use a favorable response from Rusk to "appease some members of the Board of Directors" of NBC. This suggested that strong opposition to the program remained within the network itself, in corporate suites.

The Rusk letter, drafted by Manning, barely exceeded a single page. He claimed he was "particularly grateful for your assurances that no security risk" to individuals remained "in the film as it is now edited," and he was sure it was "a moving human document." As State had long declared (in its passive-aggressive way), it remained NBC's decision on whether to air it or not, but Rusk wanted to make sure the network kept this in mind:

> You are aware, I am sure, that the principal concern of the State Department and myself was, from the beginning, the question of risks, personal and political, surrounding the covert involvement of American television or other unofficial personnel in affairs so delicate and explosive as the Berlin wall escape problem.
>
> I am sure you appreciate why such concern should exist, and why it properly should be conveyed to those who are, or might be, engaged in such undertakings in Berlin, or in other serious international situations where the material for adventurous journalism may be deeply intertwined with other considerations that affect national interests and human lives. It would be a matter of concern, too, if other enterprises were to draw from the fortunate success of the NBC operation the desire to seek similar coups—if not in Berlin in some other area where similar complications prevail.

So Rusk had eased his protest somewhat, but pointedly warned the network (again) about current and future escape enterprises.

Most revealing was the use of the word "problem" in the reference to "the Berlin wall escape problem." Perhaps the Kennedy administration's ambivalent reaction to *that*, not the NBC special, was the real "problem."

Out of the blue, NBC correspondent Sander Vanocur called Reuven Frank to tell him he had run into Attorney General Kennedy, who had brought up the tunnel controversy with alacrity. "That was a terrible thing you people did," Bobby Kennedy said, "buying that tunnel."

PERHAPS surprising even himself, Siegfried Uhse had played no role in the tunnel takedown that led to the arrest of Harry Seidel. Uhse remained well connected in escape circles, however, and continued to direct couriers for the Girrmann Group, while keeping his handler, Herr Lehmann, aware of vague plans for "border breakthroughs," as Stasi reports would inevitably describe them. Uhse's MfS pay had shot up since his second award-winning act of subversion. Until August his average monthly fee had come to only about 250 DM. That month he earned 530 DM, courtesy of the Kiefholz bust. In October, thanks to the Heidelberger tunnel bust, it was 1800 DM (or $450), an impressive sum—enough to purchase an automobile, if he wished. Uhse was instructed by his MfS bosses to "deepen relationship of trust" with Joan Glenn "to find out more about the terrorist activities of the Girrmann group." He had also managed to place a friend and fellow informer, a "Günter H," within the Girrmann community.

All of this put Uhse in prime position to learn about, and possibly expose, the new tunnel Hasso Herschel was digging under Bernauer Strasse. While it was not a Girrmann project, many of the student diggers on separate projects were now closely connected.

Herschel's excavation again opened under the swizzle stick factory, but this time in a section of the basement down the block, with a different destination in the East, on Brunnen Strasse.

Hasso had convened a meeting of veterans of the Bernauer tunnel and offered this streamlined pitch: "Well, boys, I've started a new tunnel, you want to join?" When a skeptic inquired whether NBC would be financing it, Hasso replied, "No, we don't do that anymore. I got money for that tunnel and now I can finance the new one. We have the site and this time we will do it much more perfect than last time!" He drove those who agreed to the factory straightaway. Some who fled through the original Bernauer tunnel, including Hasso's brother-in-law, Hans-Georg Moeller, also volunteered, feeling an obligation now to rescue others. Hasso tried to recruit Claus Stürmer, but Inge Stürmer vetoed this, reminding her husband that he needed to make steady money and spend more time with their two children.

Hasso's new tunnel crew had to cart away some of the dirt piled in the cellar from the previous dig to give them room to store more. Still, his second tunnel might go a little faster than the first. Equipment could be reused, and since Hasso now knew that dirt in this area was hard clay all the way to the East, he had decided to forgo wood supports.

Meanwhile, Wolf Schroedter, who still felt guilty about accepting the NBC money, took charge of another tunnel, this one started by the Girrmann Group. He broke ground with his own crew in a building at 87 Bernauer Strasse, just a few doors from the swizzle stick factory. Like Hasso, he would draw on his NBC funds—the gift that kept on giving in Berlin.

IN the last week of November, the first major magazine piece focusing on the Bernauer tunnel and related events appeared, in the *Saturday Evening Post*. It was written by veteran journalist Don Cook. The *Post* promoted the piece with small posters for newsstands reading, *Case Histories: Berlin's Tunnels to Freedom.* A similar line ran on the cover of the issue. Despite persistent claims that fifty-nine fled through the Bernauer tunnel, the article reported that it was now known as "Tunnel 29" in the "growing lore of Berlin escape stories."

Cook's lengthy article was titled "Digging a Way to Freedom," with the subhead noting, "An American television network
filmed the project—to the dismay of the State Department—
but postponed showing the film." It scrambled some of the facts
about the Bernauer tunnel—intentionally or not—and Cook
used pseudonyms for all of the key characters. Peter and Eveline
Schmidt, who were pictured with their child on a cobblestone
street in the West, were labeled "the Lohmanns." The site of the
basement was not 7 Schönholzer but 63 Werner Strasse. Others
profiled included courier "Krista" (who seems to be Ellen Schau
but with her hair color and the city where she lived changed).
Various escape helpers were quoted, always without real, or any,
names attached. One proclaimed, "Young people in Berlin are
disgusted with the Allies. . . . We are giving people hope and
saving their lives."

Gunter Lohmann (i.e., Peter Schmidt) recalled beating his
fists against a wall and crying in frustration before his family's escape. Their passage through the tunnel in September
was described in such prose as "The water leaked in, the baby
whined, and then came a burst of laughter as they raced to
safety." Cook also touched on the "controversy" around tunnel
financing. West Berlin authorities were now keeping a closer
watch on pay-for-play "and nobody is going to get away for long
with any kind of moneymaking in human lives in the city."

Also surfacing in North America was MGM's *Escape from
East Berlin*. Screened for critics in October, then shown only
in Michigan and Canada, it finally opened across the United
States to decidedly mixed reviews. Robert Siodmak's direction
was generally hailed, the screenplay much less so. Echoing some
of the reviews in West Berlin, the *Los Angeles Examiner* complained that in the film "much of the reason for wanting to leave
East Berlin concern themselves with trivia instead of principles
and true freedom." Another critic could not resist calling the
tunnel drama "shallow." At least no one slammed it as "under-
whelming."

There was, however, the added, real-life attraction of Angelika Ligma, the East Berlin student who escaped to the West under the backseat of an automobile but claimed she came through the Bernauer tunnel. Arriving in America for guest appearances, she was embellishing her already false tale with exciting fictional details for effect. *BoxOffice* headlined its brief item, "German Girl on P.A. Tour to Promote German Film." It revealed that "the girl," who recently escaped through a tunnel under the Berlin Wall, did not speak English and was conducting interviews via an interpreter. She was known only as "Fraulein Angelika" to protect her family "still in Red Germany."

ON November 30, NBC's Lester Bernstein forwarded to Robert Manning at the State Department a copy of the press release the network had readied, announcing that it had rescheduled *The Tunnel*. "I hope I'll be able to see you soon in Washington without a mission or a pressing deadline," Bernstein added.

The film was now set for telecast from 8:30 to 10:00 p.m. on Monday, December 10, six weeks after its original airdate. It was still sponsored by Gulf Oil, who had stuck with the program despite the controversy. After noting the earlier postponement, the press release declared, somewhat defensively, that "on the basis of its own judgment, NBC believes the time is appropriate to reschedule the program, which offers not only a vivid insight into one of the world's continuing major problems but an extraordinary testament to human courage and the will to be free." It quoted West Berlin officials belatedly backing the airing of the documentary.

Across the Atlantic, Reuven Frank was visiting Berlin on a trip unrelated to the tunnel controversy. When he checked in at the Kempinski Hotel he found three items awaiting him. From the Dehmel brothers he had received a full-size pick and a shovel to mark what they had gone through together. And from his home office: a telegram informing him that *The Tunnel* would finally surface.

So would Frank, who was surprised his film was airing at

all, now change his mind about leaving the network? And how many would watch his program? *The Tunnel* had been slotted against three of the most popular shows on TV, all ranked among the top eight in the Nielsen ratings: CBS comedies starring Lucille Ball, Danny Thomas, and Andy Griffith. *The Rifleman* and *Stoney Burke* would vie for viewers on ABC. In any event, Robert Kintner sent a memo to Lester Bernstein, who was about to leave the network for a top job at *Newsweek*, thanking him for evangelizing *The Tunnel* in Germany. "You handled yourself with the ingenuity of a Reston, the pomp of an Alsop, and the negotiating ability of a British prime minister," Kintner wrote. "Whether or not we get criticized—and I am sure we will to some degree—your work contributed greatly to simplifying the problem." He also wrote to Dean Rusk, thanking him for "your understanding of our position" and offering to screen *The Tunnel* for him privately if the time and date of its national airing "proves to be difficult for you."

18

Coming Up for Air

DECEMBER 1962

Three months into his six-year sentence, Hartmut Stachowitz was assigned to what was, for him, as a trained veterinarian, one of the worst jobs possible: working in the prison's pigpen, not to keep the animals healthy but to get them ready for slaughter. His wife, Gerda, who was in a separate institution, was allowed a visit by her mother, during which she finally learned that her infant son was safe and sound, living with her parents. For Manfred Meier, who received a seven-year sentence at his trial, truth—or at least one less lie—was also a consolation. He denied to the final interrogation that escape helpers had planned an armed assault at Kiefholz Strasse. This forced the Stasi to admit defeat in a note to the prosecutor and judge. "It is advised," the note read, "to ignore this contradiction."

Harry Seidel also remained in prison, awaiting a show trial set for a few days after Christmas. He still did not know who had betrayed his tunnel. (Fritz Wagner heard it was a West Berlin business owner, but didn't know who or why.) Harry did manage to get a short letter delivered to his mother, asking her to not attend his trial: "I don't want you to have pain to no purpose." The state planned to generate wide publicity, inviting numerous foreign correspondents to chronicle the proceedings at the Supreme Court of the GDR. The East German press was already in attack mode; one report claimed that Harry had instructed a tunnel comrade to kill VoPos by shooting in the shape of a cross with a submachine gun.

Seidel faced serious charges not just for the Wolfswerder plot but for all of his known escape activities, in and out of tunnels, armed and unarmed, over the previous year. The prosecution counted seven tunnel projects mounted under his leadership, four of which had been completed, leading to the escape of at least sixty East Germans. As a "violent criminal" who had "destroyed border facilities" (such as snipping barbed wire) and "endangered lives," Seidel faced at least life imprisonment. He even had to answer for attending the *Tunnel 28* premiere. The GDR's infamous minister of justice, "Red Hilde" Benjamin, had asked for the death penalty.

A Stasi interrogator told Harry snidely, *"Es geht um Kopf und Kragen"* (roughly, "Now you're in big trouble"). Harry replied, "I know how many years you gave Gengelbach. Fifteen years for me will not be enough. So there are only two other possibilities: life in prison or execution. If I can choose, I will pick the latter. Because in these times the people need a martyr."

In contrast to previous high-profile arrests, the MfS was not handing out any medals for the Wolfswerder bust. Despite the arrest of more than two dozen, including Seidel, the episode had not been widely celebrated due to the spectacular act of sabotage at its climax. The Stasi had probed but not yet solved the mystery of the tampered explosives. Investigators had interviewed or demanded written reports from all of the Stasi operatives at the scene of the bust. No one confessed to sabotage, pointed a finger at a suspect, or provided fresh evidence. Experts had determined that the wire to the explosives had been cut by a dull knife or piece of glass, then violently yanked apart, far out in the garden area of the site. It had to be deliberate, but they could not piece together who might have visited that area between the placing of the wire and the failed detonation a few hours later. Richard Schmeing, the Nazi death camp survivor who had set the wire and then, when it didn't work, inspected (by himself) a dark corner of the property, was a suspect, but investigators could find nothing to pin on him.

The MfS, meanwhile, had launched a quite different probe, into the pilfering of possessions at the Sendler cottage. The Sendlers were threatening to take their complaints to a high court in the GDR. The thievery was on such a scale that the normally stone-hearted MfS, which still suspected the couple of plotting with the tunnelers, felt it had to investigate.

ON Monday evening, December 10, between eight thirty and ten on the CBS television network, Sheriff Andy Taylor helped his deputy Barney Fife escape from a cabin where he was being held hostage by three female convicts. That madcap pair Lucy and Viv got stuck to a wall while applying heavy glue. A party for Danny Williams's son, Rusty, was wrecked by Uncle Tonoose. And a couple of channel clicks away, a very different program, featuring adventures that were far from comic, was airing at last.

More than six months had passed since Piers Anderton paid his first visit to the swizzle stick factory, three months since the successful escape, forty-one days since the program was originally slated to appear. Now *The Tunnel* had arrived, pretty much the way Reuven Frank wanted it (except for the black strips obscuring the faces of diggers and refugees who had not approved their appearance), and still sponsored by Gulf Oil. Not counting the periodic commercials, its final cut was seventy-eight minutes.

The Tunnel opened with a long shot over the Wall from a window in the NBC apartment, where (unmentioned) a white sheet once fluttered its signal for escapees across the border. The camera slowly zoomed in to find the entrance of 7 Schönholzer, with music on the sound track playing softly. It was daylight. East Berliners strolled past the entrance. Piers Anderton, in the apartment, set the scene, describing how more than two dozen refugees arrived on this street almost two months earlier.

Some had come two hundred miles. All were strangers to this place. . . . They went quietly down these cellar stairs by

a ladder down a shaft and stood fifteen feet below the surface of Schönholzer Strasse. There was a tunnel there, less
than three-feet wide and three-feet high. Through this, they
crawled . . . one hundred and forty yards to West Berlin, and
a free future. Some of the children had to be carried. I'm
Piers Anderton, NBC News, Berlin. And this is the story of
those people and that tunnel.

The documentary that followed was an aberration for prime-
time television: no interviews with participants or experts, virtually no audio beyond the narration, edited to a slow pace. It
started with the three young organizers reenacting their search
for a tunnel site. They planned the "most daring rescue operation" in Berlin's history, Anderton marveled. Then some evoca-
tive images: the Wall, the death strip, guard towers, the VoPos
"who shoot to kill." West Berliners live with this constant men-
ace but still some "decide to act." This led many to prison, others
to death.

So let's meet our heroes. There are the two Italians—the
"tall, dark and handsome" Spina and his "Sancho Panza,"
Sesta—and the blond, crew-cut Schroedter. They inspect sites
in a VW van, draw circles on maps, work out engineering equa-
tions. It's April and soon they have found the swizzle stick factory. They start chipping away at the cellar floor. Then they are
fifteen feet down. This was footage shot by the organizers before
NBC arrived on the scene.

Now the Dehmels take over as Anderton announces that
from this point all is shown "as it happened." Eight young men
are toiling "endlessly in that damp cellar." The black-and-white
footage, with primitive lighting, is crude but sharp enough for
the small screen. The sense of filth and claustrophobia are palpable. Suddenly, we are blessedly back in the bright sunlight of
Bernauer Strasse. We see a close-up of the swizzle stick factory,
disguised just enough not to be identified (a good thing, since
Hasso and friends were currently digging their next tunnel from
there). Smiling West Berliners crowd the sidewalks while their

fellow Germans, dispirited, stand in line in the East with little
to buy in stores—"life without grace . . . life merely as function."

Then back to the underground, as five tons of steel rail is
laid for the track and wooden planks are slotted on the floor. We
meet Joachim Rudolph, IDed only as *Der Kleiner*—he is "muscular" but "baby faced"—and catch a glimpse of Hasso Herschel's beard. Anderton explains how hard it was to film and
light the passage and why there was no audio at all, except for a
few seconds of footsteps or a streetcar or bus passing overhead.

Now we meet Peter and Eveline Schmidt in their modest
suburban cottage, enjoying a backyard frolic with their daughter. Clearly *The Tunnel* had been edited so that viewers, unlike
the diggers, are not stuck underground for more than a few minutes without a break. Back to Berlin: the air is growing scarce
at the digging end of the tunnel so piping must be purchased
and mounted. Then we endure the saga of the drip that became
a leak that became a flood. Now there's "mud everywhere, like
war," Anderton intones. Caked clothes are hung out to dry. A
city work crew repairs the water main, and "the flood stopped."

Sesta leaves to visit the Schmidts again in the East. We see
clippings from newspapers showing a lifeless Peter Fechter,
posted in the tunnel by angry diggers. Finally the film jumps to
September 13 and the last inspection before the escape. It's gone
according to plan—the film ignores the final leak and crucial
decision to emerge one street short of their target.

On escape day Ellen Schau, the only courier mentioned, enters the S-Bahn station and boards the train, which disappears
into the East as portentous music sounds. Then we're back to
telephoto scenes of the entrance to 7 Schönholzer. Two VoPos
walk past with a brief glance at the door. Do they know something? The sky darkens a bit. It's now six o'clock and the refugees
are on their way.

Then, no narration needed: Eveline Schmidt slowly climbs
the ladder out of the tunnel and into the West, looking shocked
to see the bright lights as Klaus Dehmel (unidentified, for obvious reasons) rushes to assist her. Her baby arrives as she nearly

collapses. Peter's mother ascends. Anderton finally speaks: "Only the faces of those who clearly consented are shown on this film. Others were edited out or where this was not possible blacked out." Hasso's sister and her child arrive. The toddlers are wiped clean by attractive young mothers with torn stockings and mud caking their legs. After three minutes of this, still with no narration, more arrive, and Anderton at last speaks again:

> To escape through a tunnel is as risky as to build one. What lies ahead is unknown. The couriers were strangers. The rendezvous could have been a trap. Death was not the greatest danger. Prison camps can be worse. Not all came. These came. These are ordinary people, not trained or accustomed to risk.

And then the key line provided by Reuven Frank's associate: "What must they be leaving to risk this?" The lengthy, moving escape sequence ends with a dramatic freeze-frame: Claus Stürmer holding his baby son for the first time.

Then a (perhaps too) quick cut to the party for diggers and escapees organized by NBC. Anderton admits there was dissension in the ranks of tunnelers, though his network is (of course) not cited as a reason—only that some had wanted to get the tunnel pumped out again so more could escape. As the camera withdraws from the party, it finds Anderton, in his trench coat, standing outside the restaurant in the dark. He tells viewers: "If East Berlin maintained its water lines better, if the tunnel had not filled with water, more dozens, perhaps hundreds of people might have been rescued. That was what the young men had hoped.

"Twenty-one of them gave half a year of their lives to dig this tunnel. But there will be other young men and other tunnels."

REUVEN Frank, who decided to stay on at NBC when the network finally scheduled *The Tunnel*, knew that it was a powerful, even

unique film, but he also knew he had made several dangerous
choices for prime-time television. When the reviews arrived it
seemed the risks had paid off.

The *Los Angeles Times* declared, "TV journalism has reached
a new and exalted plateau" with "one of the most profound and
inspiring human documents in the history of the medium." The
Boston Globe called it "probably without parallel in the brief history of television." United Press International: "A devastating
human document . . . a documentary coup of the first order." Associated Press: "Television at its best." The *San Francisco Chronicle* labeled it the "most genuinely absorbing television show of
the season." The *Detroit News*: "Seldom in its history has television presented a program with more dramatic impact." The
Pittsburgh Press: "No more forceful a statement of the difference
between life in the Free World and that under Communism ever
has been uttered." The film had not yet aired in Germany, but
reviewing the U.S. screening a writer for a major newspaper
there, the *Frankfurter Allgemeine Zeitung,* raved that "the sinister blocking of the East sector came across the TV screen as immensely impressive and far more rousing than any description
in word or picture that the U.S. has ever seen . . . resulting in an
exciting document against the inhumanity of the Wall."

Back in the United States, the White House and State Department remained pointedly silent. But watching *The Tunnel,* Bill
Moyers, the young director of the Peace Corps, felt profoundly
moved, more than by any previous film on any subject. Moyers
felt the film was like a great novel, filled with conflict, suspense,
drama, danger, struggle, and hope. But this story was true, and
it was unfolding in black and white on the small screen in his
living room. After the film ended his fingernails had left deep
marks in the heels of his hands from the sheer tension of watching. Some of the tunnelers and refugees heard a wild rumor that
President Kennedy himself had shed tears when he viewed the
program.

Expressing a decidedly minority view was the longtime
TV writer for the *New York Times,* Jack Gould, continuing the
disapproval he first expressed for the project in October. He

admitted that the film did have some "absorbing interest" but derided its "extremely slow telling"—it should have been cut to half an hour, at most an hour! Yes, escape night was interesting but "unfortunately most of the ninety-minute presentation was concerned with the mechanics of building the tunnel," which "tended to dilute the total effectiveness of the program." And he could not resist concluding with: "After seeing the program the wisdom of NBC in financially assisting the diggers still would seem questionable."

Another well-known writer, Harriet Van Horne, mocked the program for its theatricality, finding proof in the credits, which listed "Makeup by Birgitta." Reuven Frank and his team had a good laugh over that. When they had filmed the program's final scene outside the dinner party, Birgitta Anderton, fearing that the large patches of gray in her husband's beard would flash too white for the camera, had hastily eliminated the threat with her makeup pencil. As an inside joke, Frank had added "Makeup by Birgitta" to the credits.

Still, Frank was gratified with the overwhelmingly positive response. But how many viewers bothered to tune in? After several days, more sighs of relief: viewers in an estimated 18 million homes had watched, exceeding all expectations and actually matching the ratings for CBS's wildly popular sitcoms, a first for the network that year. So CBS was not only beaten to a tunnel special but also hurt by its rival's own film when it finally aired—despite CBS's tireless efforts to stop it. *The Tunnel* was nothing short of a triumph.

Reuven Frank celebrated but also recognized the role of good fortune: finding the tunnel, surviving the flood, avoiding a fatal security leak, and completing the filming with no one among his crew, the diggers, or the refugees arrested, injured, or slain. He also guessed that, in the end, the White House and State Department protests helped build viewership for his program—yet he was troubled that he still didn't understand exactly why the Kennedy administration had fought it with such vehemence. Frank realized, not for the first time but now most profoundly, how painfully vulnerable to pressure the America media remained when it came to the reporting of sensitive issues. "Anyone with half a brain," as he once put it, "can make it impossible," or nearly so.

NBC'S film was now history, but some of its stars remained busy. Wolf Schroedter's new tunnel project, opening just down the block from the swizzle stick factory, advanced well under the death strip until his diggers detected strange shoveling noises nearby. It could only mean, they concluded reluctantly, that Stasi were building their own tunnel *parallel* to Bernauer Strasse and the Wall, to cut off any passage heading East in that area. For Schroedter there would be no sequel to the September success.

This setback hardly deterred Hasso Herschel and his crew, who were pointing their shovels toward a different basement a few hundred feet past the border. All was still going well, with no one hearing any Stasi digging in the distance. Eschewing an electric rail system this time, they dragged the dirt to the tunnel's mouth in large tin butchers' pans. To minimize leaks this passage was dug much deeper than Hasso's first tunnel, almost twenty-five feet down. Even more than with the NBC tunnel, the work went nonstop; Hasso had asked the diggers to live in the factory for weeks at a stretch. Joachim Rudolph, who was dangerously behind on his university studies, was allowed to come and go, shoveling only at night and on weekends, but he always arrived for work with newspapers and magazines to distribute to his reality-deprived colleagues.

Herschel, who also continued living above ground, obtained food and drink and other supplies and delivered them to the tunnel site. Then he might join in the digging for a couple of hours. Starting in early December, and leaving the two Joachims (Rudolph and Neumann) in charge, he took brief trips with Mimmo Sesta and Gigi Spina to sell photos or NBC footage to media outlets in Paris, Zurich, Rome, and Vienna—all places where they had retained the rights.

The Wall was again hit by a series of bomb blasts. Still, West Berlin police stepped up efforts to establish friendly contacts with their counterparts across the Wall, to encourage defections or the passing of secrets. Over or through the barriers they slipped small holiday gifts, or food, cigarettes, and chocolate. The GDR guards, usually quite reserved (and no wonder), accepted the presents warmly. One West Berlin cop handed over a tub of fried chicken and watched as four guards devoured it on the spot. In thanking their fellow Germans, the GDR guards would explain that they could converse only when quite certain that none of their comrades would report them. From the exchanges that did occur, it was clear that morale in the East was plunging; most of the guards were disillusioned with their government, tired of the deprivations, "in want of everything," as one complained. Some said they wished to defect but knew their families would be punished in the aftermath.

Spirits inside the swizzle stick factory were also sagging, even with a breakthrough in sight. Some of the diggers had spent over six weeks without sunlight or loved ones, without clean clothes and a comfortable bed. They looked and smelled like what they were: tunnel rats. To get a glimpse of the outside, they would climb to a room upstairs in the factory, pull back a curtain, and gaze out at the night sky, or the rain, and grow sad or worried, pondering what friends and lovers were doing at that moment, or what they were sacrificing at school or at their jobs. Yet their dedication could not be shaken. One tunneler heard that his wife had a boyfriend and wanted a divorce. He left the site to investigate, found out this was true—and returned to the tunnel. Another digger suffered a horrible toothache, exited to have it fixed, and returned two days later, ready to pick up his shovel.

As Christmas neared, diggers were asked not to leave the factory even for the holiday, to maintain security. To ease the malaise, Herschel decided to throw a Christmas Eve party in the basement. Joachim Rudolph chose to pass up Christmas Eve with his mother and sister to attend Hasso's "office party." Herschel brought in some pine branches and stuck them in a vase to suggest a tree, plus plenty of beer, red wine, meat, and *stollen*, a seasonal bread with fruit. And a radio to play Christmas carols.

Starting at six o'clock, Hasso and his crew prepared the holiday spread. Because there weren't enough chairs for everyone, diggers dragged mattresses and sleeping bags down the steps from the makeshift bedrooms upstairs to form a circle around the "tree" and the provisions. Music wafted in the background. While this provided a much-needed respite, the mood, despite ample supplies of wine and beer, was hardly festive. Some of the bone-weary young men chatted quietly with their workmates; others, knowing what they were missing outside, said little. A few hours passed. It got to be eleven o'clock. Still Christmas Eve. The celebrants knew what they had to do; what they wanted to do. The food and bottles were packed away, mattresses returned to the sleeping quarters, the radio switched off. And then Joachim Rudolph and others on the night shift returned to their tunnel, picked up shovels and pots, and resumed their daily routine: *dig, dump, and repeat,* deeper into the East.

Epilogue

Two days after Christmas, Harry Seidel went on trial in East Berlin, just as tensions at the Wall spiked again. The third bomb blast that month tore a twenty-foot hole in the barrier and shattered at least six hundred windows in the American sector. Another East German guard escaped through barbed wire. A steel-plated bus carrying two East Berlin families rammed through the Wall under a hail of gunfire, as eight adults and children safely reached the West.

Within seventy-two hours of the start of his trial, Seidel was convicted of multiple crimes and drew a sentence of life in prison, managing to narrowly avoid the death sentence favored by the GDR's top justice official. "The extraordinary scope and dangerousness of his crimes," the judges ruled, "require permanent isolation." They likened the crimes to those committed by Nazis tried at Nuremberg. West Berlin mayor Willy Brandt, on the other hand, said there were no words strong enough to condemn this modern "inquisition" against Seidel.

Three weeks later, pressed by the Stasi in an interrogation to give up the names of the young men who completed the Kiefholz tunnel (two of them his former classmates), Seidel claimed he only knew their first names and even those "have now slipped my memory." He added, unhelpfully: "I only remember that they all had short hair." From prison Seidel wrote his wife that she should not wait for him to emerge but rather begin a new life, if she wanted. He added, "The present lack of freedom and restraint are beginning to wear on me." Far from abandoning him, she helped lead protests in Berlin that drew large crowds and international attention.

The Stasi never did find out who sabotaged the detonation of the tunnel at Wolfswerder the night Seidel was arrested. In his book *Wege durch die Mauer,* former *fluchthelfer* Burkhart Veigel makes the case for explosives expert Richard Schmeing. Veigel's analysis of the Stasi's official probe concludes that Schmeing fiddled with the detonator to prevent the explosion, then went outside and cut the cable himself to throw off any investigation.

Two months after Hasso Herschel's band of diggers spent Christmas at the swizzle stick factory, they completed their tunnel to Brunnen Strasse. On escape day the operation was busted, after a chance encounter between a friend of Hasso's sister Anita and a Stasi informer. Many were arrested, including Joachim Neumann's girlfriend, Christa. Hasso barely escaped (again).

Wolf Schroedter, after the failure of his own tunnel the previous autumn, continued to work with the Girrmann Group. When he asked the leaders where they got their hair cut, they sent him to Siegfried Uhse at a salon in the Kreuzberg district. Uhse had been "a good courier," they said. By then, Schroedter, still uncomfortable with the commercial aspects of tunnel building, had decided to sign away all of his rights to future income from sales of the NBC footage.

Uhse spied on the Girrmann office and other escape operations for another year or so. When Wolf-Dieter Sternheimer managed to get word from prison to Girrmann leaders that he believed Uhse was a Stasi agent, they replied that he seemed "harmless." The Girrmann Group soon disbanded, largely because it had been so deeply infiltrated by informers. Uhse continued his Stasi career for several years elsewhere.

Joining Uhse as an informer: Angelika Ligma, the woman who had started her career in subterfuge by posing as a tunnel escapee, a story she maintained throughout the U.S. press tour for *Escape from East Berlin*. In 1963, after a year in the West, she abandoned her new life and returned to East Germany. The Stasi told her she faced prison unless she agreed to serve as an informer. A young woman who could "make herself look good" might gain valuable confidences from "interested men," they reasoned. Ligma agreed and was given the code name "Gerda." She would work for the Stasi until 1971.

PIERS Anderton, still stewing about the Kennedy administration's attempts to kill *The Tunnel*, reached his breaking point in January 1963. The setting was NBC's annual symposium at the National Press Club. He and eight other foreign correspondents had been flown to Washington, D.C., to assess the regions they covered. It was less than a month after *The Tunnel* aired. At the luncheon Anderton found himself sitting next to Robert Manning, the chief of public affairs at the State Department. This reminded him of conflicts with State for the past year.

When Anderton got up to speak, he charged that State Department and military officials in Berlin were regularly suppressing news coverage and intimidating reporters, which prevented Americans back home from getting the full picture on Germany. He claimed that the same unnamed officials had nearly gotten him fired by leaking to *Variety* a spurious account of his speech to that women's group. State had even circulated a cable accusing him of being "pro-Communist." This smacked of the Red-baiting McCarthyism of the previous decade. Nevertheless, he vowed to return to Berlin "and fight this thing through. To move now would be to give up."

Edward R. Murrow, the former CBS newsman, approached Anderton afterward and told him, "You're my kind of guy." But NBC executives in attendance were clearly annoyed that he had spoken out in this venue. Manning confirmed the existence of a cable critical of Anderton but claimed it did not use the "pro-Communist" label. A lengthy article in *Broadcasting* magazine entitled "U.S. Censorship Overseas?" suggested that Anderton's remarks might provoke his involuntary transfer.

After a second NBC symposium in another city, Anderton wrote his wife that he was "tired of crusading" about the State Department's suppression of news. His D.C. remarks had created "a sensation," he told her. "I was called in to the State Department and reprimanded, but by this time I was angry so I fought back and made it worse," he confided. "Bill McAndrew supported me (for the time being). . . . But I really hate getting into fights, only I hate worse pussy-footing around a situation which I know is wrong."

A few days later, Anderton learned that he was being transferred to India. Due to the distance from the United States and the frequent absence of cameramen, correspondents there had difficulty getting anything on the air.

Just weeks after the State Department and White House tried to kill *The Tunnel*, the USIA, under Edward R. Murrow, ordered more than one hundred copies of the film to screen around the world. In May 1963, *The Tunnel* took home three Emmy Awards: for best documentary, for international reporting (by Piers Anderton), and for Program of the Year—the first documentary ever to earn this highest honor. Reuven Frank had carefully prepared his acceptance speech for the documentary prize, knowing he might speak to a large national audience. In it he criticized the State Department, noting that after all of its "interference" the previous October, the USIA was now "showing the film all over the world." Anderton, who had flown to New York from India, told his wife that since Frank had "blasted the State Department" in accepting his Emmy, he merely thanked the Dehmels "for winning it for me." Then, surprised with the Program of the Year award, Frank simply hailed the tunnelers— they had done all the dirty work, NBC just showed it.

Jack Gould of the *New York Times* took the opportunity to criticize NBC and *The Tunnel* once again. The Emmy wins were wholly undeserved, he wrote. The network's "subsidy" of the tunneling was "not responsible conduct. . . . That no harm was done is beside the point." Reporting the news was one thing, but "manufacturing news" to create a "drama" quite another. The Cold War "should not be a toy of show business." Gould closed by complaining that to honor "the cumbersome narration of *The Tunnel* by Piers Anderton before the sustained reportorial work of Daniel Schorr in Germany was absurd"—as if all Anderton did for *The Tunnel* was read a script.

Another columnist criticized Frank for his "bitter" remarks in accepting his Emmy.

When NBC aired a repeat of the special in August 1963 to mark the second anniversary of the Wall, *Variety* reported that the program was "SRO" on commercials. *The Tunnel* was never aired in West Germany in its entirety, however. The four diggers covered by the NBC contract sold the one-time rights to a German producer for 5000 DM. He created a version less than half the length of the original. When it aired in June 1963, some of the tunnelers held a viewing party. They unwittingly invited to it yet another Stasi informer.

PRESIDENT Kennedy's order to the CIA to begin collecting domestic intelligence on American reporters—shattering its own charter—soon was formalized as Project Mockingbird. In the spring of 1963, this resulted in the wiretapping of two columnists, Robert S. Allen and Paul Scott, after they allegedly revealed classified secrets. The source of the leak was never identified. Other reporters were also monitored in this program until its end in 1965. When declassified documents revealed Mockingbird's existence in 2007, *New York Times* reporter Tim Weiner observed, "So now the record is clear: Long before President Nixon created his 'plumbers' unit of CIA veterans to stop news leaks, President Kennedy tried to use the agency for the same goal." The *Times* separately noted: "By ordering the director of central intelligence to conduct a program of domestic surveillance, Kennedy set a precedent that Presidents Johnson, Nixon, and George W. Bush would follow."

JFK visited Berlin in June 1963 and, with Konrad Adenauer and Willy Brandt, mounted a viewing platform near Checkpoint Charlie for his first glimpse of the Wall and the death strip beyond. He was visibly moved. At City Hall he delivered his famous *"Ich bin ein Berliner"* speech before an enraptured crowd. He was assassinated five months later. Lyndon Johnson, who had resisted flying to Berlin after the Wall went up, became president. His cabinet officers Dean Rusk and Robert McNamara, along with McGeorge Bundy, promoted and defended the massive U.S. military buildup in Vietnam. Hanson Baldwin advocated an even wider escalation in articles at the *New York Times*, before retiring from the paper in 1968. The FBI probe of the man behind the July 1962 leak to Baldwin, which caused so much consternation at the White House, ended without any penalty for the leaker, Roswell Gilpatric. He later served as chairman of the Federal Reserve Bank of New York and reportedly had an affair with the widowed Jacqueline Kennedy.

The golden age of TV documentaries began to fade in 1964. The urgency to take on big issues in the Kennedy years had come to an end; so had generous budgets for nonfiction specials. "The networks appeared to have exhausted their reserves of moral courage," newsman Robert MacNeil observed. Resources had to be diverted for coverage of the Vietnam war, which in time produced its own morally courageous coverage. In any case, hard-charging reporters and highly paid anchormen were now the stars, not documentary producers like Reuven Frank.

SOON after the success of the Bernauer tunnel, courier Ellen Schau married Mimmo Sesta. Neither Sesta nor Gigi Spina worked on another tunnel after their 1962 success. Eveline Schmidt would divorce a devastated Peter Schmidt and in 1967 marry Joachim Rudolph, with whom she had danced at that party during her first weeks in the West. (Rudolph, in another twist, took trips on behalf of NBC, working as a sound man for Peter Dehmel.) The couple still have the tiny shoes her daughter Annett kicked off in the tunnel on the escape night, retrieved by Rudolph two days later.

Friedrich and Edith Sendler took the GDR to court over the items stolen from their house after the Kiefholz tunnel bust and won a rare monetary award of about $4000. They left (or were forced out of) that address, where Herr Sendler also had his carpentry workshop, within a year of the escape operation, and began living in an apartment nearby.

Joan Glenn, still working for the Girrmann Group, told Siegfried Uhse that she was obtaining a fake ID so she could act as a courier herself, and was even learning Russian. An article in an East Berlin newspaper linked her to American intelligence and falsely charged that she took part "in terrorist acts," even a bomb attack. Her Stasi file included this order: "If she enters East Berlin arrest immediately." The U.S. Mission in May 1963 summoned her to Clay Allee to warn her that she was at high risk due to her efforts to "exfiltrate" GDR citizens. She said, yes, she was "well aware" of this.

In 1964, Joachim Neumann helped dig another tunnel under Bernauer Strasse, this time from a former bakery, in an operation led by Wolfgang Fuchs, Berlin's new *tunnelmeister*. Major media in Germany funded the tunnel—the longest yet attempted, at 500 feet—in return for photo and film rights. The tunnel mistakenly broke through in a courtyard outhouse instead of a basement, but no matter: over two nights in early October, fifty-seven escaped including (at last) Neumann's girlfriend, Christa, who had just emerged from prison.

Then, East Berlin guards arrived, and one of them, Egon Schultz, was shot dead. The GDR cried murder. Tunnelers claimed that Schultz had been killed by friendly fire; the Stasi refused to release autopsy results. Tunneler Reinhard Furrer (who later became famous as a West German astronaut on the *Challenger* shuttle) was a suspect in the incident, but actually it was Christian Zobel, the angry young man identified only as "Horst P" by Flora Lewis in her October 1962 *New York Times* article, who fired a shot.

A declassified State Department cable, just days after Schultz was killed, noted "growing dissension" within the Fuchs escape organization, "which is virtually the only effective tunnel group out of some ten that once operated. Its dissolution and the circumstances of this escape could bring an end to this form of exfiltration from East Berlin." Indeed, the backlash after the death of Schultz caused West German police, government agencies, and media to drop direct support for tunnel escapes. This ended the era of escape tunnels. Only three significant tunnels were attempted in the next five years, and none would lead to a single refugee's making it to the West.

By then, Hasso Herschel and other tunnelers had become deeply involved in smuggling via automobiles. Perhaps a thousand or more East Germans came to the West hidden in Hasso's big Cadillac (often under the dashboard), in other cars and trucks, and in one case in a helicopter that Herschel wrangled. Needing money, Anita Moeller offered to drive an escape car, but her brother refused.

West German officials and business owners expanded the secret program of purchasing the freedom of prisoners in East Germany, including escape helpers Wolf-Dieter Sternheimer in 1964, Manfred Meier in 1965, and Harry Seidel in 1966. Thousands were freed this way, with the sum for each usually tied to levels of education or work skills. Manfred Meier's fiancée, Britta Bayer, for example, was "bought out" for 30,000 DM ($7500). Still, for years afterward in West Berlin, she would suffer from insomnia and bouts of paranoia. One day, hearing gunfire from the direction of the Wall, she threw herself to the floor of the apartment, frightened to death. Meier would lose a job with IBM when the American company learned that he had been imprisoned in the East—fearing, they said, that he might now be a Stasi informer. Such was the unseen reach of the Stasi.

The Stasi played one of its cruelest tricks on former courier Hartmut Stachowitz, two years after his arrest (his wife, Gerda, had already been released). He was told that he could gain his freedom if he signed a form relinquishing his West German citizenship. Desperate, he agreed, but afterward discovered that his release had already been purchased by the West. Because he had signed that paper, however, he and his family, unlike the others released, could not move to West Berlin. They did not gain permission to exit the GDR until a decade later.

After all the herculean efforts to bring her to the West, Harry Seidel's mother ended up engineering her own escape. Her mother (Harry's grandmother), ailing and ninety-one years of age, persuaded GDR officials to allow a visit to another daughter who lived in the West. Due to her frailty, the state let Harry's mother accompany her. When her visa expired, Harry's mother stayed in the West. No crawling on hands and knees necessary.

WILLY Brandt, leading the Social Democrats, became German chancellor in 1969. His aide Egon Bahr promoted a new *Ostpolitik* policy to seek limited rapprochement with the GDR. This led to some loosening of travel restrictions, with up to 40,000 East Germans a year now allowed to briefly visit the West for weddings, funerals, and other important events (the vast majority of requests, however, were refused).

The Wall, which had grown a few inches higher and thicker in most areas, remained virtually impenetrable. Escape attempts declined; deaths at the Wall fell to a handful per year. The number of Stasi agents and informers, however, only grew. Slow to warm to even modest efforts at detente, Walter Ulbricht was forced out as East German leader, replaced by the man he had put in charge of building the Wall a dozen years earlier: Erich Honecker.

Reuven Frank went on to serve two stints as NBC president. Among other innovations, he produced the network's *Weekend* series, which alternated with *Saturday Night Live* in its early seasons, and *Overnight* with Linda Ellerbee. When he retired, Robert Mulholland, who succeeded him as president of NBC, commented, "Reuven wrote the book on how the political process is covered in America." In his memoir *Out of Thin Air,* Frank revealed that it still bothered him "that no explanation made sense" for the severity of the State Department and White House response to his tunnel program. Frank filed a Freedom of Information Act request and in 1988 went so far as to contact a long-retired Dean Rusk, asking what was the real reason he was so "vigorously" against the film?

Rusk replied, "Because you were . . . endangering American interests." Thomas Schoenbaum, author of a Rusk biography, later said of him, "Yes, he certainly was involved in media suppression. He told me that is what he did with regard to Vietnam and no doubt he did the same thing with regard to Berlin. Rusk had no use for the press."

Shortly before his death, Frank was asked for a video project to defend his film one more time. Wasn't it checkbook journalism? "My best argument is: It turned out for the best," he said. "Politically, ethically, all around. The world was a better place because of it. Also, we did not enrich anybody. We carefully did not pay anybody a fee—we bought stuff." (This was not quite true, of course.) Did it bother him that he may have crossed an ethical line and that NBC itself became the focus of the story? "No," he said. "Maybe I should have a problem with that. But then it [the program] wouldn't have happened."

Piers Anderton, angry about his treatment by NBC after *The Tunnel*, relegated the display of his Emmy to the spare bathroom in his home. He made his own escape in 1964, to ABC and then to KNBC in Los Angeles, where he covered the 1968 assassination of Robert F. Kennedy, arriving on the scene just minutes after Kennedy was gunned down in the pantry of a hotel. Frustrated by the arc of his career, he retired from the news business in 1971 at the age of fifty-two. Preparing to move to England with his wife, he tossed his Emmy in the trash.

In the same period, his former rival Daniel Schorr was named to President Nixon's "enemies list." In 1976, Schorr was forced to leave the network by Richard Salant, the CBS News president (who had helped scuttle Schorr's 1962 tunnel film), after leaking contents of a secret congressional report on CIA abuses. Schorr became a prominent commentator for National Public Radio.

Schorr remained bitter about the suppression of his tunnel program until the end of his life. In an interview for the Newseum, Schorr described it as "a case of a boss of mine, who was a friend of President Kennedy, and it was possible for them [the State Department] to go to him and tell him 'the President asks you to do this.'" He paused. "And that's the story of The CBS Tunnel That Wasn't."

After delivering a lecture at Harvard, Schorr was asked by a man in the audience—his former boss, Blair Clark—about getting called off a certain unnamed story decades earlier. Like Schorr, Clark was retired; after leaving CBS, he had managed Senator Eugene McCarthy's quixotic campaign for president in 1968 and edited *The Nation*. Schorr, not missing a beat, knew exactly what Clark was coyly referring to. Schorr recalled a "wonderful" scoop about a tunnel under the Berlin Wall, and how he felt it was "malarkey" when Clark ordered him to kill it in that midnight phone call from Dean Rusk's office. Schorr added that he couldn't understand why Clark would want to mention it now at Harvard, especially since "NBC went and did it—and got prizes for having done it." Clark replied, speaking across the lecture hall, that he had no choice but to kill the film.

Meanwhile, in 1979, James P. O'Donnell, the Berlin fixer who had told Schorr about the Kiefholz tunnel, penned an improbable piece for the German edition of *Reader's Digest* predicting the fall of the Wall a decade hence and the sale of pieces of the former barrier as souvenirs. David Bowie, who lived in Berlin for several years during this period, wrote one of his most popular songs, "Heroes," after witnessing two lovers at the Wall.

In 1981, for the first time since 1962, Americans got to watch a prime-time exploration of a Berlin tunnel. In this case the vehicle was a made-for-TV drama, *Berlin Tunnel 21*—on, yes, CBS. It focused on a U.S. Army officer (Richard Thomas, the former *Waltons* star) whose German girlfriend was trapped on the other side of the Wall. "No man ever risked more for the woman he loved!" print ads for the movie declared. "A wall stands between them, but he's going to get her past the Communist troops!"

NBC's Lester Bernstein became a top editor during two

stints at *Newsweek*. James Greenfield, as a *New York Times* editor, played a key role in publication of the Pentagon Papers. His former State Department boss, Robert Manning, served for many years as the revered editor of *The Atlantic*.

Franz Baake, who helped connect the Bernauer tunnel crew to NBC, earned an Oscar nomination in 1973 for his documentary about the closing days of World War II, *Battle of Berlin*.

HARRY Seidel returned to cycling after he emerged from prison, and in 1973 won a national title with three teammates. He also held a West German government position responsible for programs that aided those persecuted under the Nazis. His former boss, Fritz Wagner, managed a popular youth hostel, then bought an inn in Bavaria, from which he reputedly made a small fortune. Seidel's friend Rainer Hildebrandt vastly expanded his collection of Wall memorabilia into a multistory museum, which remains a major tourist attraction at Checkpoint Charlie.

Joachim Neumann worked as a civil engineer on several dozen major tunnels around the world, including the passage under the English Channel connecting France and Great Britain.

Following the failure of his final tunnel in 1971, Hasso Herschel stepped away from the shrinking *fluchthelfer* community and co-owned clubs, discos, and restaurants. After they went bust, he retired to a sheep farm an hour north of Berlin. He served as chief consultant for a major German TV drama based on Tunnel 29 and received occasional checks for the sale of photos and NBC footage. Hasso's sister Anita, after nearly leaving her husband, Hans-Georg, behind in her escape to the West, remained with him and bore another child before they were divorced.

Joan Glenn, who had helped Hasso in his car-smuggling schemes, left the escape community but apparently did not return to America. The only thing known about her later activities was that she supplied the English translation for a history/

guidebook titled *In Brief Berlin*, published in 1982 by a West German federal agency.

By most accounts—and defying Stasi calculations—only about seventy-five tunnels near the Wall ever broke ground, with less than twenty judged to be successful in spiriting refugees to the West. If there were fewer escapes of any kind after the 1960s, the ones that occurred grew ever more creative. One man used a catapult to clear the Wall; another broke free with his family in a hot air balloon. (The adventure became a Disney movie.) In 1983, two men in the East shot an arrow trailing a thin nylon wire over the Wall. It landed on a rooftop in the West, where a helper strung it with steel wire, allowing the two men to slide across on pulleys. A girl in East Berlin made Soviet Army uniforms for three friends. They were saluted as they drove past the checkpoint, with her in the trunk.

IN the 1980s, new Soviet leader Mikhail Gorbachev ushered in the reforms and policies known as *glasnost*, introducing personal and economic freedoms in his nation and inspiring most of the countries behind the Iron Curtain to follow. The GDR, still led by Erich Honecker, lagged behind. Frustration and anger among East German youth threatened to boil over. The state started allowing huge rock concerts with Western stars in East Berlin as a safety valve.

When Bob Dylan was invited by a youth arm of the Communist Party to play in Treptower Park in 1987, the Stasi covered it in a six-page document filed under "Robert Zimmerman," the singer's real name. It mainly reported on logistics and security (no secret bugging of Bob, apparently). The Stasi weren't worried that Dylan would cause "undue" emotions in the crowd; he was "an old master of rock," with no particular "resonance" with the youth of the day. Dylan, in concert, more or less met their expectations.

When Bruce Springsteen headlined the following July, however, the story was quite different. The four-hour show drew his

largest crowd anywhere, perhaps 400,000, and it was beamed
to millions more via state television. Springsteen delivered an
impassioned speech in crude German: "I'm not here for any government. I've come to play rock 'n' roll for you in the hope that
one day all the barriers will be torn down." He had decided at
the last minute to change his original word, "walls," to "barriers," but the crowd went wild anyway, and the Dylan song that
followed, "Chimes of Freedom," made it all clear. Gerd Dietrich,
a German historian, later commented that Springsteen's concert
and speech "certainly contributed in a large sense to the events"
challenging the existence of the Wall. It made people "more
eager for more and more change."

Twenty-eight years after the Wall went up, East Berliners were still dying in attempts to cross it. Chris Gueffroy, age
twenty, was shot through the heart and killed one night in February 1989. As it had done from the beginning, the Stasi covered
up the real cause of death and tried to ban his funeral. When
the truth leaked out, outrage on both sides of the Wall was so
strong it forced the GDR to finally ban guards from shooting at
escapees unless their own lives were in danger. Six months later,
an electrical engineer named Winfried Freudenberg fell to his
death—in West Berlin—after he lost control of the hot air balloon that had carried him over the Wall.

These would be the final two deaths at the Wall.

The tide of history could not be resisted any longer. Neigh-
boring countries had opened their borders, and tens of thousands
of East Germans crossed into Hungary and Czechoslovakia.
Mass protests swelled in the GDR, first in Leipzig and then
Berlin. Honecker was pushed out of his leadership position. The
rock concert ploy having failed, GDR officials decided to open
another safety valve by making visas more easily available. On
the night of November 9, 1989, a government spokesman named
Schabowski went on TV to preview the new policy but bungled
the message, accidentally conveying that everyone was free to
pass through checkpoints without any approval—and that they
could do so "immediately."

Hardly believing their ears, thousands of East Germans promptly started streaming to checkpoints. At Bornholmer Strasse, more than 20,000 rushed the gates. Among them was a young chemist named Angela Merkel, an activist in the pro-government Free German Youth for many years who had lately turned against the state. Nearing midnight the guards could hold the crowds back no longer. Formalities were abandoned at other crossing points into West Berlin. Residents on both sides of the Wall were delirious, overwhelming checkpoints throughout the city. Some climbed on the Wall and danced, others smashed it with sledgehammers.

On that night, Hasso Herschel was cooking a meal in his kitchen, with the television on in the living room, when he heard the first reports. He initially could not believe them—he felt it was like a Hollywood movie unfolding. He called a few friends. "And twenty of us, thirty, even old diggers, we went to all the checkpoints and drank champagne and spent money until eleven o'clock in the morning," he would recall. "I couldn't imagine the Wall would stay open. I thought they would close it in another day or two and it would stay closed. But when that didn't happen we felt it was maybe even the end of the Cold War, and all other wars, it was our hope, our dream."

The same night, Burkhart Veigel, then an orthopedist living in Stuttgart, cried for hours in front of his TV, terribly moved. This was exactly what he had dreamed about for decades: "I wanted freedom for the people. Suddenly, they were free. It was the most important experience of my life." The next day, when his children asked him why he was still crying, he told them for the first time "what I had done back then."

A friend of Joachim Rudolph in the East had a brother living in West Berlin. The day after the Wall opened, Rudolph offered to drive him and his wife to the West to see his brother. At the border on both sides thousands of people continued to gather so it was very difficult to pass by car. Rudolph told the couple they should press their East German passports against the window and display them to people outside. When the celebrants in the streets saw this, they burst into cheers and knocked on the car roof in approval—"an amazing situation," Rudolph later said.

During the following days and then weeks, police on both sides began to remove parts of the Wall to build more border crossing points, at Brandenburg Gate, Potsdamer Platz, and elsewhere. "Very often I was there to watch it," Rudolph said. "Many cars with satellite dishes and reporters were there, and many Berliners came to watch. I remember in that time this terrible weather, but I was there at night many hours with an umbrella—and next morning I had to go to work. In my life I never will forget that exciting time."

Crowds of East Berliners ransacked Stasi headquarters, then secured rooms with files holding hundreds of millions of pages. Countless other documents had been shredded by Stasi staffers in their final days there, until the shredders burned out from overuse. Over 170,000 Stasi informers would be identified by name in the files—about 10,000 of them under the age of eighteen—but estimates of the actual number went as high as half a million, and even much higher if occasional collaborators were included.

Special reports in the days after the fall of the Wall drew weak TV ratings in America, however. "It just didn't play," said one network spokesman. "Everyone who has touched this story saw their ratings go down," reported an ABC producer. Reuven Frank opined: "Maybe if four people sat around and talked about it with Oprah Winfrey, people would listen."

After Erich Mielke, the despotic MfS chief since the 1950s, was finally forced to resign, he was swiftly sent to prison. He would be convicted of committing two murders going back to the 1930s, but a separate trial charging him with ordering the shooting of refugees at the Wall was terminated when it became clear that he was mentally unfit (which, one might say, had always been the case).

Much of the fabled barrier that ran through the center of the city was torn down within months, as if the scar and the symbolism could not be banished fast enough. Germany's full reunification occurred on October 3, 1990, less than one year after the opening of the Wall. Berlin again became one city. The historian Fritz Stern called the era of reunification "Germany's second chance."

Two years later, a German magazine arranged a conversation between Harry Seidel and Dr. Heinrich Toeplitz, the chief GDR judge who had directed his trial and helped sentence him to life imprisonment almost thirty years earlier. Toeplitz refused to apologize. Seidel said, "Nevertheless, I wish you well."

SIEGFRIED Uhse, for unknown reasons, became less enthusiastic about working for the Stasi in the late 1960s, failing to complete assignments to spy on the French military in Baden-Baden and the CIA in Berlin. Perhaps feeling some measure of guilt, he volunteered at Amnesty International in an effort to get his former Stasi colleague "Günter H" and several political prisoners out of the GDR. The MfS severed their relationship in 1977.

After the Wall fell, Hartmut and Gerda Stachowitz tracked Uhse's address to Berlin and tried to persuade the authorities to bring him to trial. Officials replied that his crimes were too distant. A former GDR associate claims he took a stroll along Kiefholz Strasse with Uhse in 2004. Uhse, he says, told him that he betrayed the tunnel there only to help further "world peace," but still the former agent "Hardy" couldn't help but think about the dozens of fellow Germans he earmarked for prison. Uhse reportedly died in Thailand a few years later.

Unlike Uhse, many GDR soldiers who shot escapees were arrested after unification. In nearly every case—including the two charged with shooting Peter Fechter—they were convicted, but released and sentenced to probation. At their trial in 1997, two guards who fired at Fechter said they were sorry but claimed they had not intended to kill him. Fechter had remained the most famous martyr of the Cold War in Berlin. Numerous books, plays, and songs were written about him. "You can draw a direct line from the moment of Peter Fechter to the moment where the

smaller part, the oppressed part, of Germany collapses," Egon
Bahr observed. In the West, a memorial cross and garden directly across the Wall from where he fell drew visitors by the
thousands for many years before being leveled in urban renewal.
A small but striking stone monument, often with flowers at its
base left by visitors, stands today on a sidewalk near the same
spot, but most people pass it without a glance.

Christian Zobel, accused by many, in both East and West,
of firing the shot in 1964 that killed GDR guard Egon Schultz,
would live a tortured life, drink heavily, and die before he turned
fifty. Finally, in 1992, an official inquiry sustained the original
claim by escape helpers, finding that the bullet that killed the
guard actually came from the gun of one of Schultz's colleagues.

Germans who had been arrested—or believed they had been
spied on—by the MfS were allowed exclusive access to their
Stasi files. Often they discovered that friends, neighbors, or even
family members had monitored them or divulged information
that led to their arrests. One former GDR border guard, Ulrich
Mühe, had become a stage and film actor, and a leader of some
of the protests in 1989 that led to the fall of the Wall. When he
obtained his Stasi file, he learned that his own wife had been a
Stasi informer. They divorced. He went on to play a conflicted
Stasi operative (who ultimately does one good deed) in the movie
The Lives of Others, which would win the Academy Award for
Best Foreign Language Film in 2006.

By the turn of the century, what was called *ostalgie,* or nos-
talgia, for the GDR era was growing in its former territory. A
2009 poll of eastern Germans for *Der Spiegel* found that about
half believed that criticism of the former state was overblown
and agreed with the statement that it had "more good sides
than bad sides—there were problems but life was good." A historian named Stefan Wolle said these "rose-tinted" responses
showed that many believed "the value of their own history is
at stake." Political scientist Klaus Schroeder explained, "Many
eastern Germans perceive all criticism of the system as a personal attack."

Of the Wall itself, Anna Funder observed in her acclaimed book *Stasiland* that most people in the former East Germany "want to forget it. In fact, it seems now most people on both sides want to pretend it was never there. The Wall has been erased so quickly that there is hardly a trace of it in the streets." When Berlin marked the twenty-fifth anniversary of the fall of the Wall in November 2014, thousands of lights had to be temporarily installed to trace its former route. The anniversary drew wide media coverage around the world. NBC chose this moment to post *The Tunnel* in its entirety on its website. An accompanying article pegged the network's total payments to the Bernauer tunnelers in 1962 as about $150,000 in current value.

Under Chancellor Angela Merkel, Germany remains one of America's close allies but its citizens, according to opinion polls, harbor deep ambivalence about the United States. To a significant extent, the country is still divided politically, with a surprising level of anti-democratic feeling (and opposition to new immigrants) in the former East and plenty of left-wing sentiment in the old West. Peter Schneider, a well-known German journalist and author (one of his books is *The Wall Jumper*) told a *New Yorker* writer that Americans in the Cold War era "created a model of a savior, and now we find by looking at you that you are not perfect at all—much less, you are actually corrupt, you are terrible businessmen, you have no ideals anymore."

IN an October 2012 ceremony, Harry Seidel and thirteen other ex-tunnelers and couriers finally received one of Germany's highest awards, the Federal Cross of Merit. Two of them, Joachim Rudolph and Hasso Herschel, now lead tours of tunnel sites for the Berliner Unterwelten. This popular organization has published several valuable books on the escape era. It also built, underground, a replica of the opening yards of Tunnel 29, complete with steel rail, buckets, and carts, just a few blocks from where it actually started on Bernauer Strasse.

Several blocks nearby have been transformed into one of

the world's most informative and moving memorial sites, with a
museum, a guard tower, sections of the Wall, and several walk-
able acres of the former death strip. Photographs of all who died
attempting to escape, including Heinz Jercha, Siegfried Noffke,
and Peter Fechter, are displayed outdoors at the Window of Remembrance, embedded side by side in a rusty steel wall just
yards from remnants of the original concrete one.

While the Wall recedes into history, heated debate contin-
ues over controversial barriers elsewhere, from the U.S.-built
fence and wall in the American Southwest (which some politicians, pundits, and citizens wish to extend to cover the entire
2000-mile-long border with Mexico) to the so-called peace walls
marking Catholic and Protestant enclaves in Northern Ireland.
The massive concrete barrier Israel has built along, and then inside, the West Bank beginning in 2002 is now nearly three times
as long, and in many places twice as high, as its Berlin fore-
bear. With its array of checkpoints, electric fencing, guard towers, patrol roads, and death strips, it strongly evokes the Berlin
Wall in its heyday. The International Court of Justice, Amnesty
International, Human Rights Watch, and the World Council of
Churches have all condemned the existence of the wall, or its
route destroying Palestinian farmland, or both. "The wall is a
symbol that we cannot live together," an Israeli told his friend
David Hare during one of the playwright's visits to Tel Aviv. "It's
an admission of failure."

In Berlin today, construction cranes break the skyline almost everywhere one looks, to the east or to the west. That is
yet another legacy of the Wall, as its removal "left vast empty
spaces that have turned out to be civic boons in ways politicians
and planners twenty-five years ago did not foresee," a *New York
Times* reporter judged. Berlin remains one of the hippest and
trendiest cities of the new century.

The swizzle stick factory at Bernauer Strasse, the former
Sendler cottage on Kiefholz Strasse, and most of the other starting and exit points for Berlin's vintage tunnels have long been
demolished. The battered old tenement at 7 Schönholzer Strasse

still stands, however, now offering luxury housing following a recent renovation. A listing for an elegant apartment just across the street described a setting vastly changed from half a century earlier, claiming a location in "one of Berlin's most vanguard neighborhoods: the recently dubbed New Art District," with "cutting-edge art galleries" and "a wonderful choice of cafés, restaurants and pubs filled with young modern types, in true Berlin fashion."

The only hint of history on this block is a large plaque at the entrance to 7 Schönholzer, erected to the left of the door in 2009. "The tunnel," it explains, "was dug by brave men who chose this dangerous way so that they could embrace again their wives, children, relatives and friends" who were trapped "across the inhumane border."

When Joachim Rudolph happened to visit Schönholzer Strasse and saw that the tenement was undergoing that high-end renovation, he feared that the vintage white enamel plate over the door bearing the black numeral "7" would soon be replaced and discarded. That "7" meant nothing to the workmen, but it had special meaning for him, especially since he was the tunneler who had ventured from cellar to street on September 14, 1962, to make sure his crew had broken through at the right address—and spotted this black numeral, on this plate, to his profound relief. Now, he feared, the marker might disappear forever.

Returning late that night with a few tools, Rudolph (who was pushing seventy) carefully climbed the scaffolding and removed his prize. Later, far to the west in the Charlottenburg district, he mounted the plate on one of the doors inside his comfortable walk-up apartment, where this German citizen who once dug a 400-foot tunnel under the Berlin Wall, and the woman who was first to crawl through it, now admire their lucky 7 every day, and know what it means to them, and what it represents for the rest of the world.

ACKNOWLEDGMENTS

When I started this book, I had no idea how many of the key tunnelers, couriers, and escapees were still living, or if they would be reachable, and if so, would be willing to speak with me. I made a list and, surprisingly, I was able to arrange lengthy interviews with nearly all of them, and then some. A hearty bunch! So thanks, first of all, to those who made extra time for interviews and follow-ups: Hasso Herschel, Joachim Neumann, Uli Pfeifer, Wolf Schroedter, Harry Seidel, Claus Stürmer, Boris Franzke, Ellen Sesta, Hartmut and Gerda Stachowitz, Manfred Meier, Eveline (Schmidt) Rudolph, and Anita Moeller.

On the NBC side, I received valuable insight (and many photographs) from Piers Anderton's widow, Birgitta Anderton; Reuven Frank's sons, Jim and Peter Frank; and Lester Bernstein's son, Paul Bernstein, and daughter, Nina Bernstein. Jim Greenfield provided essential insights into the thinking and actions of Dean Rusk at the State Department and communications with the Kennedy White House.

Three of the foremost experts on escapes at the Wall offered vital assistance on historical facts and public attitudes. Dr. Burkhart Veigel did this on a frequent basis and shared critical documents. I would also like to thank Dietmar Arnold (who helped provide photos) at Berliner Unterwelten, and Maria Nooke. Their books are also often cited in the notes.

Several archivists provided significant source material, but I would like to single out Stacey Chandler at the John F. Kennedy Presidential Library, who returned again and again to collections there, and repeatedly found surprising and significant cables, memos, and letters. I was warned that, due to strict rules on confidentiality, I would be lucky to receive much that was new or valuable from the Stasi (BStU) archives in Berlin, especially since I needed a response quickly. Instead, I gained access, via Annett Müller, to hundreds of pages of Stasi reports on the 1962 tunnels, and on many of the key players, many of which few or no one had seen before. Meanwhile, researcher Satu Haase-Webb tackled for me the massive, and often baffling, State Department and CIA files at the National Archives in College Park, Maryland.

My agent, Gary Morris (at the David Black Agency), took on the proposal for this book at a very early stage, before I'd ever met him, and provided not only encouragement and enthusiasm but vital advice on revising key portions of it. Brian Siberell (at Creative Artists Agency) was able to sell the proposal for a possible movie. From the start I had hoped that Rachel Klayman at Crown would end up editing the book, based on my experience with her on two previous projects. My wish came true and, as expected, she offered brilliant guidance (with a soft touch) throughout the months that followed. Her associate Meghan Houser provided close editing of the manuscript at several stages that was quite exceptional. I'd write more about these two but they made me promise to keep the acknowledgments short!

And now, five people to whom I owe special thanks.

This book was inspired by my first trip to Berlin shortly after my daughter, Jeni Mitchell, her husband, Stephane Henaut, and their four-year-old son, Jules, moved there. As fate would have it, they live little more than a mile from Bernauer Strasse, in the former East Berlin. After I began work on this book, Jeni, who had just earned her Ph.D., got me off and running: forwarding or summarizing articles from journals, advising on sources, and setting up our visit to the Stasi archives and the first request for files.

Since I barely speak a word of German, one can imagine the language difficulties I faced. Stephane, who is half-German, therefore provided a number of absolutely critical services: helping arrange and then accompanying me on all interviews in Berlin, and later transcribing tapes; translating many of the Stasi documents and sections of various books; and even conducting a few key interviews on his own in my absence. He was extraordinary at all of this and also a witty sidekick.

To my wife, Barbara Bedway, I offer heartfelt thanks for not only sharing many of my experiences in Berlin but also encouraging me to pursue this subject, and then reading and improving three separate versions of the manuscript. She also drafted most of those extraordinary sections in the book about Peter Fechter.

Emely von Oest, an actress (and former airline pilot) living in Los Angeles, grew up in East Berlin during the Wall's middle to late period. I came to know her as one of the "stars" of the documentary I coproduced on Beethoven's Ninth Symphony. Like Stephane, she translated a vast number of Stasi documents and portions of books, and even conducted a couple of phone interviews (while leading a fruitless search for Siegfried Uhse). For more than a year she offered almost daily advice or research help from her unique, personal perspective. I can't thank her enough.

Finally, there is Joachim Rudolph. He not only endured almost ten hours of interviews but also put me in touch with most of the other tunnelers, and some of the couriers and escapees—the value of that contribution can hardly be overstated. He analyzed Stasi documents, and even got the notoriously reticent Harry Seidel to answer several dozen questions. And, like Emely, he became an almost daily source of advice and fact checking, making sure my own "digging" would not spring a leak and helping me avoid even a partial narrative collapse. Joachim, I could not have finished this book in anything close to this form without you. Your nickname might be *Der Kleiner*, but in contributing to this book you have been more like *Der Gigant*.

NOTES

The following abbreviations are used in the notes:

BStU/MfS   Files from Stasi archives (Der Bundesbeauftragte für die Unterlagen des Staatssicherheitsdienstes), Berlin, Germany

int(s)   Interview(s) conducted by or on behalf of the author

JFK-BC   Berlin Cables, State Department cables to and from Berlin, Bonn, and Washington, D.C., National Security Files, John F. Kennedy Presidential Library, Boston, Massachusetts

JFK-NSF   National Security Files, John F. Kennedy Presidential Library, Boston, Massachusetts

JFK-PDB   Daily CIA briefing pages, released 2015, see http://www.foia.cia/collection.PDBs

JFK-POF   President's Office Files, John F. Kennedy Presidential Library, Boston, Massachusetts

JFKL   Oral histories and other material at John F. Kennedy Presidential Library, Boston, Massachusetts

MPA   Motion Picture Academy, Margaret Herrick Library, Los Angeles, California, files for *Escape from East Berlin/Tunnel 28*

NARA-CIA   National Archives and Records Administration, College Park, Maryland (CIA internal weekly and periodic intelligence reports)

NARA-De   National Archives and Records Administration, College Park, Maryland (State Department and CIA documents released in January 2014 by the National Declassification Center)

Newseum   Newseum, Washington, D.C. (video interviews and transcripts of outtakes tied to exhibit on the Berlin Wall)

NYT   *The New York Times*

PR of JFK   Timothy Naftali, Philip Zelikow, and Ernest May. *The Presidential Recordings of John F. Kennedy: The Great Crises.* 3 vols. New York: W. W. Norton, 2001

RFP   Reuven Frank Papers, Tufts University, Medford, Massachusetts

1. THE CYCLIST

1   **Harry Seidel loved action**: For Harry Seidel and his background see Harry Seidel int.; Pierre Galante, *The Berlin Wall* (Garden City, N.Y.: Doubleday, 1963), 1–138; Burkhart Veigel, *Wege durch die Mauer* (Berlin: Berliner Unterwelten, 2013), 225–45; Dietmar Arnold and Sven Felix Kellerhoff, *Die Fluchttunnel von Berlin* (Berlin: Propylaen, 2008), 114–23.

3   **On the evening of September 3**: Galante, *Berlin Wall*, 98–103.

4   **Since shortly after World War II**: Frederick Taylor, *The Berlin Wall* (London: Bloomsbury, 2006), 18–185; Frederick Kempe, *Berlin 1961* (New York: G. P. Putnam's Sons, 2011), 3–362; W. R. Smyser, *Kennedy and the Berlin Wall* (Lanham, Md.: Rowman & Littlefield, 2010), 1–123; Peter Wyden, *Wall* (New York: Simon & Schuster, 1989), 1–219.

5   **As the summit ended**: Kempe, *Berlin 1961*, 257–61; Smyser, *Kennedy and the Berlin Wall*, 71–75.

6   **"Next week, if they chose to close"**: Kempe, *Berlin 1961*, 315; Smyser, *Kennedy and the Berlin Wall*, 90.

7   **President Kennedy said nothing**: Kempe, *Berlin 1961*, 316–18; Smyser, *Kennedy and the Berlin Wall*, 89–95.

7   **When Secretary of State Dean Rusk**: Smyser, *Kennedy and the Berlin Wall*, 105; Wyden, *Wall*, 177.

7   **"The East Germans have done us"**: Smyser, *Kennedy and the Berlin Wall*, 108.

7   **East Berlin was an armed camp**: Wyden, *Wall*, 161; Kempe, *Berlin 1961*, 368.

7   **Edward R. Murrow**: Taylor, *Berlin Wall*, 217–18; Smyser, *Kennedy and the Berlin Wall*, 115–16.

8   **"It's not a very nice solution"**: Taylor, *Berlin Wall*, 220.

8   **Kennedy's CIA briefing on August 14**: JFK-PDB.

8   **Berlin Task Force**: Taylor, *Berlin Wall*, 226–27.

9   **"that bastard from Berlin"**: Ibid.

9   **jumping out windows**: For the initial series of deaths at the Wall, see Hans-Herman Hertle and Maria Nooke, eds., *The Victims at the Berlin Wall* (Berlin: Ch Links, 2011), 36–56; Kempe, *Berlin 1961*, 363–66; Pertti Ahonen, *Death at the Berlin Wall* (New York: Oxford University Press, 2011), 32–37.

10   **A Berlin sculptor**: John Bainbridge, "Die Mauer: The Early Days of the Berlin Wall," *The New Yorker*, Oct. 27, 1962.

11   **"We are sold"**: Joseph Wechsberg, "Letter from Berlin," *The New Yorker*, May 26, 1962.

11   **Harry Seidel was so angry**: Galante, *Berlin Wall*, 136–38, 168.

12   **One of Harry's fellow escape artists**: Hertle and Nooke, *Victims at the Berlin Wall*, 62–64.

12   **aiding the Ministry for State Security**: For background on the Stasi, see John O. Koehler, *Stasi: The Untold Story* (Boulder, Colo.: Westview Press, 1999), 10–220.

13   **Seidel met Fritz Wagner**: Galante, *Berlin Wall*, 144–50; Seidel int.

14   **When U.S. attorney general Robert Kennedy**: David E. Murphy, Sergei A. Kondrashev, and George Bailey, *Battleground Berlin* (New Haven, Conn.: Yale University Press, 1997), 388.

16   **Harry pushed his luck too far**: My account of the death of Heinz Jercha is drawn from Seidel int.; Hertle and Nooke, *Victims at the Berlin Wall*, 74–76; Veigel, *Wege durch die Mauer*, 109–24; Arnold and Kellerhoff, *Die Fluchttunnel von Berlin*, 114–24; files on this operation, BStU/MfS, HA I Nr 6086.

18   **Another young *fluchthelfer***: Burkhart Veigel int.

18   **"East German guards shot and killed"**: "Foreign Students Aided Escape of 600 East Berliners to West," *NYT*, March 28, 1962.

19   **"But that's my job"**: Galante, *Berlin Wall*, 152.

2. TWO ITALIANS AND A GERMAN

20   **Among those taking heart**: Ellen Sesta, *Der Tunnel in die Freiheit* (Munich: Ullstein, 2001), 11–49; Arnold and Kellerhoff, *Die Fluchttunnel von Berlin*, 214–31; Ellen (Schau) Sesta, Wolf Schroedter ints.; *Der Tunnel*, documentary directed by Marcus Vetter (1999).

20   **arts student named Peter Schmidt**: Eveline (Schmidt) Rudolph int.

22   **his boss back at 30 Rockefeller Plaza**: Background on Reuven Frank is from his *Out of Thin Air* (New York: Simon & Schuster, 1991), 7–192; *NYT*, Feb. 7, 2006.

23   **A native of San Francisco**: Obituary, *NYT*, Sept. 23, 2004; Birgitta Anderton, Kit Anderton, Mary Anderton ints.

24   **On at least one occasion**: Piers Anderton to Reuven Frank, n.d. (c. 1988), RFP.

25   **"In adventurous ways"**: "Escape Through the Wall," *Der Spiegel*, March 1962 (issue 13).

26   **"Scarlet Pimpernel raids"**: "Foreign Students Aided Escape of 600 East Berliners to West," *NYT*, March 28, 1962.

27   **Uhse first visited**: BStU/MfS, Uhse file, March 20, 1962.

28   **Uhse had served as**: BStU/MfS, Uhse file, Aug. 21, 1963; Veigel, *Wege durch die Mauer*, 260–65.

29   **"I, Siegfried Uhse"**: BStU/MfS, Uhse file, Sept. 30, 1961.

29   **"Uhse is sure that"**: BStU/MfS, Uhse file, March 20, 1962.

29   **They didn't yet have funding**: Schroedter, Sesta ints.; NBC, *The Tunnel*.

30   **They located the owner**: Schroedter int.

3. THE RECRUITS

32   **Reuven Frank had posted**: Jim Frank, Peter Frank ints.

32   **It was the golden age**: A. William Bluem, *Documentary in American Television* (New York: Hastings House, 1965), 89–144; Mary Ann Watson, *The Expanding Vista* (New York: Oxford University Press, 1990), 135–52.

32   **CBS's man in Berlin**: Daniel Schorr, *Staying Tuned* (New York: Washington Square Press, 2001), 147–81; Wyden, *Wall*, 161.

33   **"Small numbers of East Germans"**: Wyden, *Wall*, 220–21; Schorr, *Staying Tuned*, 155–60.

33   **Schorr provided riveting coverage**: Wyden, *Wall*, 260.

34   **Schorr had come to believe**: Daniel Schorr interview, Newseum.

34   **Like Piers Anderton**: Schorr, *Staying Tuned*, 161–63.

35   **Clark had told Schorr**: Ibid., 165.

35   **MGM announced that it would**: *Hollywood Reporter*, Feb. 27, 1962; *Variety*, March 27 and April 12, 1962.

36   **Through the Student Union**: Schroedter int.

37   **The adoptive mother of their friend**: Sesta, *Tunnel in die Freiheit*, 42–47; Sesta, Schroedter ints.; *Der Tunnel*, Vetter.

37   **Joachim Rudolph and Manfred Krebs**: Schroedter int.; Joachim Rudolph int.

38   **On a bike tour**: J. Rudolph int.

38   **A woman who lived on a high floor**: Kempe, *Berlin 1961*, 394.

38   **Philip Held**: Hertle and Nooke, *Victims at the Berlin Wall*, 77.

38   **A nine-year-old East Berlin boy**: "East Berlin Boy, 9, Leaps to Safety," *NYT*, April 11, 1962.

39   **Three young East Berliners**: Hertle and Nooke, *Victims at the Berlin Wall*, 80.

39   **Horst Frank**: Ibid., 86.

39   **"Defectors will not be permitted"**: U.S. Mission in Berlin to State Dept., June 16, 1962, JFK-BC.

39   **After any fatal shooting**: Hertle and Nooke, *Victims at the Berlin Wall*, 23–25.

40   **Dozens of border guards**: Ibid., 83.

41   **"there is no evidence"**: CIA file, NIE #12.4-62, May 9, 1962, NARA-CIA.

41   **Despite the tensions along the Wall**: Permanent exhibit at Wall Memorial (*Gedenkstatte Berliner Mauer*), www.berliner-mauer-gedenkstaette.de, Berlin.

42   **Anyone in the East**: Joachim Neumann int.

42   **Indeed, Siegfried Uhse's Stasi career**: BStU/MfS, Uhse file, March 20 and April 4, 1962.

44   **The three students had found**: Schroedter, J. Rudolph ints.

44   **Several days later, two men who had heard**: Hasso Herschel, Uli Pfeifer ints.

45   **The new recruits had met**: Herschel, Pfeifer ints.; Anita Moeller int.; Hasso Herschel interview, Newseum.

46   **He had a concrete goal**: Herschel, Moeller ints.

4. THE PRESIDENT

48   **His press conferences**: Watson, *Expanding Vista*, 72–74; Pierre Salinger, *P.S.* (New York: St. Martin's Press, 1995), 94–95.

48   **"Well, I am reading more"**: John F. Kennedy, press conference transcript, May 8, 1962; Salinger, *P.S.*, 96.

49   **It seemed to Sorensen**: *USA Today*, Nov. 20, 2013.

49   **To his friend Ben Bradlee**: Benjamin C. Bradlee, *Conversations with Kennedy* (New York: W. W. Norton, 1975).

49   **the American Newspaper Publishers Association**: Amy L. Heyse and Katie L. Gibson, "John F. Kennedy: 'The President and the Press,'" *VOD Journal*, vol. 9, available at http://voicesofdemocracy.umd.edu/wp-content/uploads/2015/11/JFK-Revision_Essay_Final-lph-edits.pdf.

50   **"Tension along the Wall"**: Briefing of May 30, 1962, available at www.foia.cia/collection.PDBs.

51   **"may lead to serious"**: Ibid., May 21.

51   **the Stasi was masterly**: Koehler, *Stasi: The Untold Story*, 1–142.

53   **After several days of digging vertically**: J. Rudolph, Schroedter, Pfeifer, Herschel ints.

54   **Fortunately, several other young adventurers**: J. Neumann int.

54   **One day in May the pair visited**: Ibid.

55   **Max Thomas**: Christopher Hilton, *The Wall* (London: Sutton, 2001), 180.

56   **"Grandpa Fritz"**: "Eastberliner, 81, Tunnels to West," *NYT*, May 19, 1962.

56   **One shooting caused**: Taylor, *Berlin Wall*, 307–9; Ahonen, *Death at the Berlin Wall*, 66–83.

57   **"Under no circumstances should you"**: Taylor, *Berlin Wall*, 309.

57   **Yet the shootings continued**: Hertle and Nooke, *Victims at the Berlin Wall*, 89.

58   **A second, if far less intimidating**: Associated Press, May 12, 1962.

58   **MGM hoped the movie**: Wyden, *Wall*, 160.

59   **The director, however, had trouble finding**: MGM press kit, 1962, MPA.

59   **"scarcely contain the excitement"**: *Los Angeles Times*, May 21, 1962.

59 **On another occasion, Siodmak**: MGM press kit, MPA.

60 **When Franz Baake responded**: Fritz Baake int.; Schroedter int.; *Der Tunnel*, Vetter.

61 **Meyer, who was working**: Fritjof Meyer int.; Baake int.; *Der Tunnel*, Vetter.

61 **Abe Ashkenasi**: Meyer int.; Abe Ashkenasi int.; Piers Anderton to Reuven Frank, July 24, 1988, RFP.

62 **he observed the manager, Bodo Köhler**: BStU/MfS, Uhse file, May 7, 1962.

62 **Joan Glenn had come to Stuttgart**: Two articles in Stanford University's newspaper, the *Stanford Daily*, on Jan. 4, 1961, and Feb. 11, 1962, were valuable sources for Glenn's background.

62 **Berlin had become a kind of mecca**: Veigel, *Wege durch die Mauer*, 17–202; Taylor, *Berlin Wall*, 305–6; Neumann int.

63 **fellow Stanford exchange students**: *Stanford Daily*, Feb. 11 and May 16, 1962.

63 **Joan Glenn asked Siegfried Uhse**: BStU/MfS, Uhse file, May 23, 1962.

63 **early morning hours of May 26**: "Four Blasts in 15 Minutes Rip Reds' Wall in Berlin," *NYT*, May 27, 1962.

64 **Hans-Joachim Lazai**: Kempe, *Berlin 1961*, 392–93; Wall Memorial (*Gedenkstaette Berliner Mauer*), Contemporary Witnesses, www.berliner-mauer-gedenkstaette.de.

64 **The leaders of**: Veigel int.

65 **Bodo Köhler had met with Egon Bahr**: Ibid.

65 **After they discussed taking stronger measures**: *Handelsblatt*, June 12, 2007.

66 **On May 27**: Baake, Meyer ints.; Anderton journal, May 27–28, 1962, courtesy of B. Anderton.

66 **Anderton arrived to find**: NBC, *The Tunnel*.

67 **Anderton told Schroedter**: Anderton to Frank, July 24, 1988, RFP.

67 **Schroedter made him sign an agreement**: Schroedter personal archive, courtesy of B. Anderton.

5. THE CORRESPONDENT

68   **a brief civil ceremony**: B. Anderton int.

68   **Reuven Frank was enjoying**: Frank, *Out of Thin Air,* 193–94;
Reuven Frank, "Making of *The Tunnel,*" *Television Quarterly*
(Fall 1963), 11–12; Anderton to Frank, July 24, 1988, RFP.

68   **When the party ended**: B. Anderton int.

69   **organizers had demanded $50,000**: Frank, *Out of Thin Air,*
193–94; Anderton to Frank, July 24, 1988, RFP; Reuven Frank
interview, "Emmy TV Legends," Archive of American Television, http://www.emmytvlegends.org/.

69   **Anderton hastily arranged a pair**: B. Anderton int.

70   **That night**: Ibid.

70   **His troubles arrived**: Anderton to Frank, July 24, 1988, RFP.

70   **a "very negative"**: Allen Lightner to Dean Rusk, April 2, 1962,
and Dean Rusk to Allen Lightner, April 2, 1962, JFK-BC.

71   **"has been highly critical"**: George Ball to State Department,
April 6, 1962, JFK-BC.

71   **Anderton had drawn more criticism**: "NBC's Anderton's Incendiary Berlin Talk Shocks Wives of U.S. VIPs," *Variety,*
April 18, 1962.

71   **a State Department hit job**: B. Anderton int.; *Variety,* Aug. 29,
1962.

71   **As the Bernauer tunnel**: Galante, *Berlin Wall,* 159–66; Arnold
and Kellerhoff, *Die Fluchttunnel von Berlin,* 214–31; Seidel int.

72   **A Stasi report called the pub**: Veigel, *Wege durch die Mauer,*
257–69.

73   **Above ground on Heidelberger**: Ulf Mann, *Tunnelfluchten*
(Berlin: Transit Buchverlag, 2005).

73   **Two more women arrived with children**: Galante, *Berlin Wall,*
165.

74   **"East German officials"**: "Two Groups Escape from East Berlin," *NYT,* June 12, 1962.

74   **When Harry Seidel returned home**: Galante, *Berlin Wall,* 170.

74   **"The wall has not fulfilled"**: "Two Groups Escape from East
Berlin."

75   **Axel Hannemann**: Hertle and Nooke, *Victims at the Berlin Wall,*
91.

75   **a group of fourteen East Berliners**: Sydney Gruson, "Red Guard Killed at Berlin's Wall," *NYT*, June 19, 1962.

76   **The clash was triggered**: Taylor, *Berlin Wall*, 317–18; Ahonen, *Death at the Berlin Wall*, 83–93; Hilton, *The Wall*, 184–85; Sydney Gruson, "Clashes at Wall Feared as Desperation Grows in East Berlin," *NYT*, June 20, 1962.

76   **"[w]armer weather and the summer"**: "Current Intelligence Weekly Summary," June 29, 1962, NARA-CIA.

77   **Piers Anderton, it turned out**: Baake, Meyer ints.

78   **Mimmo Sesta warned them**: Sesta, *Tunnel in die Freiheit*, 69–75.

78   **The NBC contract**: Schroedter personal archives.

78   **One reason the tunnel organizers**: Schroedter and J. Rudolph ints.

79   **Soon the tunnelers were directly below**: J. Rudolph, Pfeifer, Neumann, Herschel ints.

80   **historic Kempinski Hotel**: B. Anderton int.

81   **acquiring additional firearms**: Sesta, *Tunnel in die Freiheit*, 49–75.

81   **He wondered whether he should have**: Frank, "Making of *The Tunnel*," 12.

82   **Frank had mandated**: Ibid., 12–13; Frank, *Out of Thin Air*, 194–95.

82   **June 20 would be**: Anderton journal, courtesy of B. Anderton; Anderton to Frank, July 24, 1988, RFP; Frank, *Out of Thin Air*, 198.

83   **His June 15 report**: Transcripts of the three reports are in Daniel Schorr Papers, Library of Congress, Washington, D.C.

84   **a native of Georgia**: For Dean Rusk's background, see Thomas J. Schoenbaum, *Waging Peace and War* (New York: Simon & Schuster, 1988); Smyser, *Kennedy and the Berlin Wall*, 26–29.

85   **"imprisoned"**: Smyser, *Kennedy and the Berlin Wall*, 86, which is also the source for details of Rusk's Berlin trip.

85   **Uhse's latest visit**: BStU/MfS, Uhse file, June 23, June 25, June 27, June 30, 1962, reports.

86   **a tidy three-man operation**: Hertle and Nooke, *Victims at the Berlin Wall*, 99; Taylor, *Berlin Wall*, 317.

6. THE LEAKS

88   **The schedule for one week**: Joachim Rudolph provided me
      with a copy of this document.

88   **The Dehmels tried to capture**: NBC, *The Tunnel*.

88   **The camera might introduce**: Outtakes from NBC's *The Tun-
      nel*, provided by J. Rudolph.

89   **the musky smell of centuries past**: NBC, *The Tunnel*.

89   **When they received the invoice**: These invoices are in Wolf
      Schroedter personal archive.

89   **eagerly displayed a machine gun**: J. Rudolph, Herschel ints.

89   **The organizers decided to tell two more**: J. Rudolph, Schroedter
      ints.

90   **Sesta and Spina visited**: Schroedter int.; Sesta, *Der Tunnel in
      die Freiheit*, 77–86; *Der Tunnel*, Vetter.

91   **Known to activists**: Schroedter int.

91   **Mertens who first put Bodo Köhler**: Veigel, Schroedter ints.

91   **So Spina and Sesta trotted off**: *Der Tunnel*, Vetter.

92   **name, address, and age of everyone**: Ibid.

92   **As a handful of city workers**: NBC, *The Tunnel*; Pfeifer int.
      Mimmo Sesta believed something quite different occurred:
      that the West Berlin workers actually just turned off the water
      going to the East (and into the tunnel) and merely faked doing
      repairs to mislead the GDR guards (Sesta, *Der Tunnel in die
      Freiheit*).

93   **At least the diggers could rest**: NBC, *The Tunnel*.

93   **"I saw and heard what happened"**: Ibid.

93   **They drove around the city**: Schroedter in ibid.

94   **Shirley MacLaine**: Schorr, *Staying Tuned*, 166–67.

95   **a Silver Bear medal**: Baake int.

95   **The order had come directly**: *PR of JFK*, 1:xviii–xix.

96   **From East Berlin came news**: "Reds Sentence American
      Youth," *NYT*, June 16, 1962; "Berlin Reds Tell of Tunnel
      Fight," *NYT*, July 7, 1962.

96   **"show trial"**: "East German Show Trial Sends Five Men to
      Prison," *NYT*, July 5, 1962.

97   **The perpetually impatient**: Seidel int.

97   **Mertens told Dieter Thieme**: Schroedter int.

98   **On July 27**: Schroedter, J. Rudolph, Seidel ints.

99   **Siegfried Uhse's diligence was paying off**: BStU/MfS, Uhse file, July 20, 1962.

99   **Glenn ordered Uhse**: BStU/MfS, Uhse file, July 27, 1962.

100 **He was to go East**: BStU/MfS, Uhse file, July 28, 1962.

100 **It didn't take him long to inaugurate**: *PR of JFK*, 1:4.

100 **"There's nothing else we have to do"**: Ibid., 1:45.

101 **a nearly two-hour discussion**: Ibid., 1:80–89.

101 **Steuart L. Pittman**: "U.S. Offers Facts to Back Shelters," *NYT,* May 12, 1962.

101 **The front-page *Times* article**: Hanson Baldwin, "Soviet Missiles Protected in Hardened Position," *NYT,* July 26, 1962.

102 **The White House quickly moved**: Robert B. Davies, *Baldwin of the "Times"* (Annapolis, Md.: Naval Institute Press, 2011), 265; *PR of JFK*, 1:18.

102 **When Baldwin answered his door**: Davies, *Baldwin of the "Times,"* 264–67.

102 **Baldwin received a call**: Transcript, Baldwin investigation file, JFK-NSF.

103 **J. Edgar Hoover notified**: Davies, *Baldwin of the "Times,"* 267–68.

103 **James P. O'Donnell**: Wyden, *Wall,* 172–75; Kempe, *Berlin 1961,* 313–15.

104 **"There'll be a lot of shooting"**: Kempe, *Berlin 1961,* 384.

104 **Now, in late July**: Ralph A. Brown, "For the Record" (memo), Aug. 3, 1962, NARA-De.

7. SCHORR AND THE SECRETARY

105 **The meeting between Daniel Schorr**: Brown, "For the Record," memo, Aug. 3, 1962, NARA-De.

106 **O'Donnell asked staffer Ralph A. Brown**: Ibid.

106 **"I expressed concern"**: Ibid.

107 **"We consider it would be"**: Arthur Day to Percival, Aug. 3, 1962, NARA-De.

107 **"They naturally approve your attempt"**: Percival to Arthur Day, Aug. 3, 1962, NARA-De.

107 **Hulick broached the subject**: Charles Hulick to Hillenbrand, Aug. 4, 1962, NARA-De.

108 **Jim O'Donnell informed the Mission**: Ibid.

108 **Hulick cabled a report**: Ibid.

108 **Unsure if Harry**: J. Rudolph int.

108 **an evening meeting convened**: BStU/MfS 3733/65 44.

109 **JFK had approved**: *PR of JFK*, 1:188–201.

109 **"the judgment of your board"**: Central Intelligence Briefing, July 28, 1962, NARA-CIA.

111 **It was the kind of sensitive issue Dean Rusk**: James L. Greenfield int.; Salinger, *P.S.*, 92–93.

112 **Rusk briefed Greenfield**: Greenfield int.

112 **"being compromised"**: Ibid.

112 **Anderton broke Reuven Frank's**: Reuven Frank to Lester Bernstein, Feb. 2, 1988, RFP.

113 **Wolf-Dieter Sternheimer**: Wolf-Dieter Sternheimer, Renate Sternheimer, Manfred Meier, Britta Bayer ints.

113 **scout the Kiefholz area**: Meier int.

114 **After August 6**: BStU/MfS, Uhse file, Aug. 7, 1962; W.-D. Sternheimer int.

115 **One of the first to learn**: Moeller int.

116 **At 10:50 p.m.**: Dean Rusk to Charles Hulick, Aug. 7, 1962, NARA-De; Rusk appointment book, Dean Rusk Papers, Lyndon B. Johnson Presidential Library, Austin, Texas; Greenfield int.

116 **Born in 1917**: Obituary, *NYT*, June 8, 2000; David Halberstam, *The Powers That Be* (New York: Knopf, 1979), 384.

117 **Jacqueline Kennedy's guided tour**: Watson, *Expanding Vista*, 139–44.

117 **Now Dean Rusk was asking**: Greenfield int.; Schorr, *Staying Tuned*, 165.

117 **When Schorr got a call**: Berlin Brigade, U.S. military log, Aug. 7, 1962, NARA, College Park, Maryland.

117 **"What's this I hear"**: Schorr, *Staying Tuned*, 165–66; Schorr int., Newseum.

118 **The call lasted just six minutes**: Berlin Brigade, U.S. military log, Aug. 7, 1962.

118 **Dean Rusk finally got a chance**: Rusk to Hulick, Aug. 7, 1962, NARA-De.

8. KIEFHOLZ STRASSE

120 **Here he was, in West Berlin**: BStU/MfS, Uhse file, Aug. 7, 1962, and Uhse notes; W.-D. Sternheimer int.; map: Veigel, *Wege durch die Mauer*, 269.

121 **He learned that the escape**: BStU/MfS, Uhse file, Aug. 7, 1962.

122 **Uhse told his handlers**: Ibid.

122 **Anita Moeller and her husband**: Moeller int.

122 **The Schmidts had received word**: E. Rudolph int. She believes that they did not have time to contact her husband's mother. However, Ellen Sesta claims in her book that Mimmo spotted the mother, with the Schmidts, at a truck rendezvous point. If true, they may have been on the truck that turned around, and did not walk to Puder Strasse.

123 **Elsewhere in East Berlin**: Interviews with Hartmut and Gerda Stachowitz, BStU/MfS, 3733/65 44.

123 **When Stachowitz arrived**: Interview with Hartmut Stachowitz interview, BStU/MfS, Uhse file, Aug. 7, 1962.

124 **Mimmo Sesta had decided**: Sesta, *Der Tunnel in die Freiheit*, 99–107.

124 **Manfred Meier was growing nervous**: Meier int.

124 **Thanks to Siegfried Uhse's report**: For the mobilization in the area of the Sendler house, see BStU/MfS, 3733/65; BStU/MfS, HAI 13256; BStU/MfS, BV Berlin Abt VII nr. 1553.

125 **Pacing up and down the street**: Meier int.

126 **At 5:15 p.m., two other**: For this and other details, see BStU/MfS, 3733/65; BStU/MfS, HAI 13256; BStU/MfS, BV Berlin Abt VII nr. 1553.

126 **the three-man breakthrough team**: Herschel, J. Rudolph, Pfeifer, Schroedter, Joachim Neumann ints.

128 **News of the pending tunnel escape**: Memo, Lieutenant Colonel Gerald Sabatino, Aug. 7, 1962, NARA-De.

128 **Lightner cabled Secretary of State Rusk**: Allen Lightner to Dean Rusk, Aug. 7, 1962, NARA-De.

129 **When Hartmut and Gerda Stachowitz**: Interviews with Hartmut and Gerda Stachowitz, BStU/MfS, 3733/65.

129   **The Moeller and Schmidt**: Moeller, E. Rudolph ints.

130   **Also fleeing the scene**: Interview with Gengelbach, BStU/ MfS, 3733/65 44, and BV Berlin Abt VII nr. 1553 19.

130   **Mimmo Sesta, meanwhile**: Sesta, *Der Tunnel in die Freiheit*, 99–107; Sesta int.

130   **After his arrest he was marched**: Meier int.

131   **Hasso Herschel had no idea**: Herschel, J. Rudolph, Pfeifer ints.

133   **Across the border fence**: NBC footage, Aug. 7, 1962, NBC Universal archives.

134   **The three diggers, unaware**: Herschel, Pfeifer, J. Rudolph, Neumann ints.

136   **When they climbed out of their cavern**: Pfeifer, Herschel, J. Rudolph ints.

137   **arrested at the Heinrich-Heine checkpoint**: W.-D. Sternheimer int.

137   **The Stasi wasted no time**: Interviews with Friedrich Sendler and Edith Sendler, BStU/MfS, 3733/65.

138   **Faring even worse**: Gerda Stachowitz int.

9. PRISONERS AND PROTESTERS

139   **Secretary of State Rusk did not hear**: Allen Lightner to Dean Rusk, Aug. 7, 1962, NARA-De.

139   **Rusk received no further word**: Allen Lightner to Dean Rusk, Aug. 8, 1962, NARA-De.

140   **Also on August 8**: "Memorandum for the Record," Aug. 8, 1962, NARA-De.

140   **An embassy staffer in Bonn**: Unidentified staffer to Allen Lightner, Aug. 8, 1962, NARA-De.

140   **"US officials have no apology"**: Dean Rusk to U.S. Mission in Berlin, Aug. 10, 1962, NARA-De.

141   **After inspecting the tunnel**: BStU/MfS, HAI 13256 59, 98, and 100.

142   **Manfred Meier, meanwhile**: Meier int.; BStU/MfS, 3733/65.

142   **She sat on the edge**: Bayer int.

142   **Sure enough, the interrogation**: BStU/MfS, 3733/65.

143   **the Girrmann Group remained perplexed**: BStU/MfS, BV Berlin Abt VII nr. 1553 19; BStU/MfS, Uhse file, Aug. 8, 1962.

143 **Siegfried Uhse was summoned**: BStU/MfS, Uhse file, Aug. 9, 1962.

145 **"I have been informed"**: BStU/MfS, Uhse file, Aug. 9, 1962.

145 **On the afternoon of August 10**: Morris to Dean Rusk and Allen Lightner, Aug. 10, 1962, NARA-De.

146 **Schorr visited Lightner**: Allen Lightner to Dean Rusk, Aug. 12, 1962, NARA-De.

146 **U.S.–NATO nuclear strategy**: Kempe, *Berlin 1961*, 299–305, 431–36.

147 **A top Strategic Air Command**: Ibid., 435.

147 **A U.S. target list**: *NYT,* Dec. 22, 2015.

147 **When his brother had met him**: Kempe, *Berlin 1961*, 263.

148 **Henry Kissinger**: Ibid., 302–4.

148 **Now Kennedy wanted to know**: *PR of JFK*, 1:31–330.

150 **He was the only son**: For Fechter background, see Lars Broder Keil and Sven Felix Kellerhoff, *Mord an der Mauer* (Berlin, Bastei Lubbe, 2012), 1–28; *Ein Tag im August: Der Fall Peter Fechter,* film directed by Wolfgang Schoen (2012); Hertle and Nooke, *Victims at the Berlin Wall,* 102–3.

151 **"and with minimum possible opportunity"**: Charles Hulick to Dean Rusk, Aug. 12, 1962, NARA-De.

151 **"for precisely what"**: Central Intelligence Briefing, Aug. 11, 1962, JFK-PDB.

152 **Another action signaled**: Schroedter, J. Rudolph ints.

152 **the duty officer's log**: Berlin Brigade, U.S. military log, Aug. 13, 1962, NARA.

153 **a chaotic scene**: BStU/MfS, HAI 13256.

10. THE INTRUDER

155 **Even as Herschel**: Herschel, J. Rudolph, Pfeifer, Neumann ints.

156 **Then another tunneler found out**: Herschel int.; Sesta, *Tunnel in die Freiheit,* 77–97; *Der Tunnel,* Vetter; Hasso Herschel interview, Newseum.

157 **"about a Berlin matter"**: Robert Kintner to Dean Rusk, Nov. 20, 1962, RFP.

157 **The meeting, which had been cleared**: Ibid.; Greenfield int.

158   **Greenfield felt it had gone**: Greenfield int.

158   **he saw an arm squeezing through**: J. Rudolph, Herschel ints.;
       Claus Stürmer int.

162   **"Through reports and information"**: BStU/MfS, 13337/64.

162   **another internal report**: BStU/MfS, 3733/65.

163   **After a few such sessions**: W.-D. Sternheimer int.

163   **Manfred Meier had once believed**: Meier int.

163   **Also undergoing heavy**: BStU/MfS, 3733/65 86.

164   **Then one or both**: BStU/MfS, HAI 13256 16, 59, and 71.

165   **August 17 dawned gray**: The account of the Fechter shooting
       is based on *NYT,* Aug. 18, 1962; Hertle and Nooke, *Victims at
       the Berlin Wall,* 102–5; Keil and Kellerhoff, *Mord an der Mauer,*
       29–50; Ahonen, *Death at the Berlin Wall,* 54–63; Berlin Bri-
       gade, U.S. military log, Aug. 17, 1962, NARA; Hilton, *The Wall,*
       189–99; *Ein Tag im August,* Schoen.

166   **Renate Haase**: *Bild Zeitung,* July 22, 2012.

167   **Bypassing the normal chain**: Wyden, *Wall,* 273.

168   **One of those who heard**: Margit Hosseini, "The Wall," epi-
       sode 9, www2.gwu.edu/~nsarchiv/colwar/interviews. An
       American student from Stanford, Dennis Bark, rushed to the
       scene from Checkpoint Charlie. See Hilton, *The Wall,* 190–92.

168   **Wolfgang Bera**: Keil and Kellerhoff, *Mord an der Mauer,* 57–
       59; Christoph Hemann, "Snapshot and Icon," February 2005,
       www.zeithistorische-forschungen.de/2-2005/id%3D4512.

168   **Also in the neighborhood**: Keil and Kellerhoff, *Mord an der
       Mauer,* 50–59.

169   **"The matter has taken care"**: Wyden, *Wall,* 274.

169   **By late afternoon**: The account of the reaction and rioting is
       based on Berlin Brigade logs, Aug. 19–21, 1962; *NYT,* Aug. 19–
       21, 1962.

170   **Dieter Breitenborn**: Keil and Kellerhoff, *Mord an der Mauer,*
       57–59, 77.

11. THE MARTYR

172   **An official morning-after autopsy**: *Ein Tag im August,* Schoen.

172   **Violent demonstrations**: Ahonen, *Death at the Berlin Wall,* 55–
       56; Keil and Kellerhoff, *Mord an der Mauer,* 80–81.

173   **"This is going to have"**: Arthur Day int.

173   **At 8:40 that morning**: Keil and Kellerhoff, *Mord an der Mauer*, 113.

173   **The young men digging**: Neumann, Pfeifer, J. Rudolph ints.

174   **The official East German report**: *Neues Deutschland*, Aug. 18, 1962; Hilton, *The Wall*, 194–96; "Report of the Commander of 1st Border Brigade," www.chronik-der-mauer.de.

174   **Unlike many others arrested**: BStU/MfS, 3733/65 44; BStU/MfS, 18284/63 261; BStU/MfS, 1596/64 105.

174   **Whenever Gerda asked**: G. Stachowitz int.

175   **One turn of events**: G. Stachowitz, H. Stachowitz ints.

175   **Manfred Meier, also imprisoned**: Meier int.

175   **composed a detailed scenario**: BStU/MfS, Meier file.

176   **his civilian clothes**: Meier int.

177   **Its political chief**: Arthur Day memo, Aug. 19, 1962, Office of the Historian, vol. 15, Berlin Crisis, Document 95, Department of State, Washington, D.C.

177   **More troubling**: Day cable and transcript of conversation, Aug. 19, 1962, JFK-NSF.

178   **a peeved Dean Rusk**: Dean Rusk to Charles Hulick, JFK-BC.

178   **a rather colloquial memo**: Chester V. Clifton memo, Aug. 20, 1962, JFK-NSF.

178   **When the ambulance**: Charles Hulick to Dean Rusk, n.d., JFK-BC.

178   **Claus Stürmer had bolstered**: C. Stürmer int.

179   **The time off allowed**: Schroedter int.

180   **a partial retraction**: "Clarify Piers Anderton Speech," *Variety*, Aug. 29, 1962.

180   **Their only rendezvous**: Frank, *Out of Thin Air*, 194.

181   **Frank, who had not visited Berlin**: Ibid., 199.

181   **"While I recognize"**: John F. Kennedy to Dean Rusk, Aug. 21, 1962, JFK-NSF.

181   **Not willing to wait**: JFK-POF.

182   **Adlai Stevenson**: *PR of JFK*, 1:552–65.

182   **the leak exploited by Hanson Baldwin**: Ibid., 1:595–99.

182   **The tap on**: Davies, *Baldwin of the "Times,"* 265–67.

183   **The meeting opened:** *PR of JFK*, 1:593–603.

185   **"The American Commandant was informed":** George Ball to John F. Kennedy, Aug. 24, 1962, JFK-NSF.

185   **Hans-Dieter Wesa:** Hertle and Nooke, *Victims at the Berlin Wall*, 106–8.

185   **"My Dear Erika":** Keil and Kellerhoff, *Mord an der Mauer*, 95.

186   **Ten days after that murder:** AP, Aug. 26, 1962; UPI, Aug. 27, 1962; Reuters, Aug. 27, 1962.

187   **"The newsmen were accused":** Berlin Mission to Bonn embassy, Aug. 20, 1962, JFK-BC.

187   **In America:** "The Boy Who Died on the Wall," *Life*, Aug. 31, 1962.

187   **Its Luce-owned sibling:** "Wall of Shame," *Time*, Aug. 31, 1962.

188   **"I think the truth":** John F. Kennedy to Mike Mansfield, Aug. 28, 1962, JFK-NSF.

188   **Rusk, Bundy, and others:** *PR of JFK*, 1:627–40.

188   **"strange and frightening week":** Sydney Gruson, "City's Mood: Anger and Frustration," *NYT*, Aug. 26, 1962.

12. COMING UP SHORT

190   **The show trial:** *Neues Deutschland*, Aug. 30–Sept. 5, 1962.

191   **Frau Sendler seemed deeply distressed:** W.-D. Sternheimer int.

192   **On the evening of their release:** BStU/MfS, HAI 13256.

193   **she attempted to smuggle:** BStU/MfS, 3733/65.

194   **Those interrogating Manfred Meier:** Meier int.

195   **That afternoon Mundt rode:** Hertle and Nooke, *Victims at the Berlin Wall*, 109–11.

196   **the digging went on as usual:** Schroedter, Herschel, J. Rudolph ints.; Sesta, *Der Tunnel in die Freiheit*, 151–55.

197   **An American spy plane:** *PR of JFK*, 2:4–20.

198   **take up the Cuba question:** Ibid., 2:19–50.

199   **Premier Khrushchev:** Smyser, *Kennedy and the Berlin Wall*, 189.

200   **Plotting how far they had:** J. Rudolph, Pfeifer, Neumann ints.

200   **Joachim Neumann had one other:** Neumann int.

201   **They did alert NBC**: Frank, *Out of Thin Air*, 196; Frank, "Making of *The Tunnel*," 16; Piers Anderton to Reuven Frank, July 24, 1988, RFP; Harry Thoess to Reuven Frank, Jan. 21, 1988, RFP.

201   **he rented an apartment**: Sesta, *Der Tunnel in die Freiheit*, 156–59.

202   **What they had to concentrate on**: Herschel, Pfeifer ints.

202   **This was Ellen Schau**: Sesta int.; Sesta, *Der Tunnel in die Freiheit*, 175–93.

203   **Meanwhile, Stindt and Anderton**: J. Rudolph int.

203   **Ellen and Mimmo shared breakfast**: Sesta int.; Sesta, *Der Tunnel in die Freiheit*, 185–97.

204   **From the NBC apartment**: NBC, *The Tunnel*.

204   **Anita Moeller had not spoken**: Moeller int.

205   **a tram to Bernauer Strasse**: Sesta, *Der Tunnel in die Freiheit*, 175–93; Sesta int.

205   **preparing for the breakthrough**: Herschel, J. Rudolph, Pfeifer ints.

206   **Christian Bahner**: Schroedter, J. Rudolph, Herschel ints.

207   **Reuven Frank had received**: Frank, *Out of Thin Air*, 196; Frank, "Making of *The Tunnel*," 16–17.

13. SCHÖNHOLZER STRASSE

209   **Joachim Rudolph brought**: J. Rudolph, Herschel, Sesta, Moeller ints.

210   **Back at the NBC office**: Frank, *Out of Thin Air*, 196; Frank, "Making of *The Tunnel*," 16–17.

210   **Peter Dehmel**: NBC, *The Tunnel*; Sesta, *Der Tunnel in die Freiheit*, 195–222.

210   **Once the courier left**: E. Rudolph int.; *Der Tunnel*, Vetter. Eveline believes that Ellen Schau was this courier, but she is not certain and Schau vigorously denies this.

211   **By then, Ellen Schau**: Sesta int.; Sesta, *Der Tunnel in die Freiheit*, 189–240; Frank, "Making of *The Tunnel*," 17.

211   **the four-man breakthrough crew**: Herschel, J. Rudolph, Pfeifer ints. Thomas Bahner says that his brother Christian Bahner (deceased) claimed to have helped raise money for the tunnel and to have aided it in other ways long before arriving for the breakthrough, but this could not be confirmed (Bahner int.).

212   **Imagine the shock and alarm**: Neumann, Pfeifer ints.

213   **"I'm willing to risk going"**: Pfeifer int.

213   **Reaching the far end**: The account of the breakthrough and escape action is drawn from Herschel, J. Rudolph, Neumann ints.

214   **Hasso, with his very rare**: Herschel, J. Rudolph ints.

216   **arrived separately at the bar**: E. Rudolph, Moeller ints.

217   **Then a slender young woman**: Sesta, *Der Tunnel in die Freiheit*, 202–22; Sesta, E. Rudolph, Moeller ints.

218   **the two groups of Schmidts**: E. Rudolph, Herschel, J. Rudolph ints.

219   **A woman's pocketbook**: NBC, *The Tunnel*; E. Rudolph int.

220   **Anita Moeller was looking**: Moeller, Herschel, Neumann ints.

220   **Emerging at the other end**: NBC, *The Tunnel*; Moeller int.

221   **nervously completing her rounds**: Sesta int.

222   **Ellen grabbed a taxi**: Ibid.; Sesta, *Der Tunnel in die Freiheit*, 231.

222   **Inge Stürmer had decided**: Inge Stürmer int.; *Bild Zeitung*, Jan. 16, 2001.

223   **organize her two children**: *Der Tunnel*, Vetter; I. Stürmer int.

223   **He had a list of refugees**: Herschel int.; *Der Tunnel*, Vetter.

224   **The two Joachims**: Neumann, J. Rudolph ints. Rudolph recalls Neumann coming out into the back of the lobby with a gun to "cover" him, but Neumann says he remained by the door to the cellar, ready to act.

225   **middle of the tunnel**: Pfeifer int.

225   **At last the refugees**: Neumann, J. Rudolph ints.

226   **At the other end**: I. Stürmer, C. Stürmer ints.; NBC, *The Tunnel*.

226   **Not waiting for**: Neumann, J. Rudolph ints.

227   **At the NBC office**: Frank, *Out of Thin Air*, 197; Frank, "Making of *The Tunnel*," 16–17; Piers Anderton to Reuven Frank, July 24, 1988, RFP.

14. UNDERGROUND FILM

228   **On the morning after**: Neumann, Pfeifer, C. Stürmer ints.; *Der Tunnel*, Vetter.

229   **"We got them all out"**: B. Anderton int.

229   **While Anderton and**: Frank, *Out of Thin Air*, 197; Frank, "Making of *The Tunnel*," 17; Piers Anderton to Reuven Frank, July 24, 1988, RFP; *The Tunnel*, NBC.

230   **West Berlin's newest citizens**: Moeller, I. Stürmer, E. Rudolph ints.

231   **Angelika Ligma**: BStU/MfS, 7413/72.

231   **The plan to spring more refugees**: Neumann int.

232   **meeting at the swizzle stick factory**: Schroedter, J. Rudolph, Pfeifer ints.

233   **two tiny shoes**: J. Rudolph int.

233   **until September 18**: "Twenty-Nine East Berliners Flee Through 400-Foot Tunnel," *NYT*, Sept. 18, 1962; *Washington Post*, Sept. 19, 1962.

234   **"ultimately close to sixty"**: Charles Hulick to Dean Rusk, Sept. 19, 1962, RFP.

235   **"The flight from the GDR"**: BStU/MfS, 2743/69 189.

235   **Bodo Köhler even gave Uhse**: BStU/MfS, Uhse file, Sept. 29, 1962.

236   **One East German who**: BStU/MfS, 7413/72.

237   **a child's doll**: B. Anderton int.

237   **a week or two editing**: Frank, *Out of Thin Air*, 197–98; Frank, "Making of *The Tunnel*," 17–20.

238   **Würzburger Hofbräu**: *The Tunnel*, NBC; Moeller, J. Rudolph ints.

239   **Then Piers Anderton stepped outside**: *The Tunnel*, NBC.

239   **threw themselves another party**: J. Rudolph int.

239   **Peter Schmidt, watching them**: *Der Tunnel*, Vetter.

240   **Rabbits roaming the death strip**: UPI, July 23, 1962.

240   **Reports filled the logs**: Berlin Brigade, U.S. military logs, Sept. 1962, NARA.

241   **"U.S. press officers"**: Charles Hulick to Dean Rusk, Sept. 22, 1962, JFK-BC.

241   **"show by example"**: McGeorge Bundy to John F. Kennedy, Sept. 20, 1962, JFK-NSF.

241   **"a careful line"**: McGeorge Bundy to Henry Kissinger, Sept. 14, 1962, JFK-NSF.

242   **On September 28**: "Current Intelligence Weekly Summary," Sept. 28, 1962, NARA-CIA.

242   **a proponent of pictures over words**: Bluem, *Documentary in American Television*, 105–6, 218–19.

244   **Afraid to check their cargo**: Frank, *Out of Thin Air*, 202; Frank, "Making of *The Tunnel*," 21.

15. THREATS

245   **Frank felt immensely relieved**: Frank, *Out of Thin Air*, 203–5.

245   **An article with no byline**: "Tunnels Inc.," *Time*, Oct. 5, 1962; Val Adams, "TV Film Records Refugees' Flight," *NYT*, Oct. 5, 1962.

246   **"the only disagreeable development"**: Piers Anderton to Birgitta Anderton, n.d. [c. Oct. 1962], courtesy of B. Anderton.

246   **a remarkable October 5 cable**: George Ball to U.S. Berlin Mission, Oct. 5, 1962, RFP.

247   **"Reported Anderton enterprise"**: Charles Hulick to Dean Rusk, Oct. 6, 1962, JFK-BC.

248   **"blow up"**: Greenfield int.

248   **Since walking away**: Galante, *Berlin Wall*, 190–93; Seidel int.; Veigel, *Wege durch die Mauer*, 274–84.

248   **the Stasi was more alert**: Veigel, *Wege durch die Mauer*, 274–83; BStU/MfS, Uhse file, Oct. 6, 1962.

249   **At daybreak, before the Stasi**: BStU/MfS, ZKG 7917; Veigel, *Wege durch die Mauer*, 276–80; Galante, *Berlin Wall*, 190–93; BStU/MfS, Uhse file, Oct. 6, 1962.

250   **his wife's fury**: Galante, *Berlin Wall*, 193.

250   **"U.S. and British instructing police"**: Charles Hulick to Dean Rusk, Oct. 6, 1962, JFK-BC.

250   **On October 10**: Robert Kintner to Dean Rusk, Nov. 20, 1962, RFP.

251   **Frank confirmed details**: Frank, *Out of Thin Air*, 202.

251   **"while they were sorry"**: Robert Manning to Dean Rusk, Oct. 11, 1962, RFP.

252   **"Officials are wondering"**: Max Frankel, "Cold War Confusion," *NYT,* Oct. 13, 1962.

252   **"There has been much more"**: Piers Anderton to Birgitta Anderton, Oct. 14, 1962, courtesy of B. Anderton.

252   **"quick and courageous action"**: BStU/MfS, Uhse file, 13337/64.

253   **its handwritten log**: BStU/MfS, HA I, 4300.

253   **Stanford University radio station**: *Stanford Daily,* Oct. 11, 1962.

254   **Frank screened a rough cut**: Frank, "Making of *The Tunnel,*" 20.

254   **demand by Hasso Herschel**: Herschel int.; Hasso Herschel interview, Newseum; contract in Schroedter personal archives.

255   **Peter Schmidt managed**: J. Rudolph int.

255   **His wife, Eveline**: E. Rudolph int.

255   **"to coordinate possible new approach"**: Allen Lightner to Dean Rusk, Oct. 15, 1962, RFP.

256   **Lester Bernstein**: Reuven Frank to Lester Bernstein, Feb. 2, [1988], RFP; Nina Bernstein int.

256   **rang a dorm at TU**: Herschel int.

256   **"thoughtless, business-minded capitalism"**: *Arbeiter Zeitung* (Vienna), Oct. 13, 1962.

256   **"every consideration"**: Robert Manning to William McAndrew, Oct. 17, 1962, RFP.

257   **The conversation was civil**: Statement by William McAndrew, n.d., RFP.

258   **a delicate relationship**: Elie Abel oral history, JFKL.

258   **"This is a program"**: Frank, *Out of Thin Air,* 210; Reuven Frank interview, Newseum; Rusk appointment book, Dean Rusk Papers, Lyndon Baines Johnson Presidential Library, Austin, Texas.

259   **"If we solve the Berlin problem"**: David G. Coleman, *The Fourteenth Day* (New York: W. W. Norton, 2012), 177; Theodore Sorensen, *Kennedy* (New York: Harper & Row, 1965), 669.

259   **When October 16 dawned**: Kempe, *Berlin 1961,* 495–98; *PR of JFK,* 2:391–468.

259   **On his notepad**: "Doodles" file, JFK Personal Papers, JFKL.

260   **Robert Kennedy handed**: *Newsweek,* Aug. 13, 2 000.

260   **Sacrificing a few Cuban missiles**: Ibid., 2:573.

262   **Near midnight**: Ibid., 2:576–77.

16. BURIED TUNNEL

263 **Frank sought refuge**: Frank, "Making of *The Tunnel*," 21.

263 **two pages of "guidance"**: Oct. 12 and Oct. 18, 1962, RFP.

264 **"a community of information officers"**: Pierre Salinger oral history, JFKL.

264 **Sure enough, the first**: Transcript, Robert J. Manning Papers, Yale University, New Haven, Connecticut.

265 **"misunderstanding"**: Lincoln White statement, Oct. 19, 1962, RFP.

265 **Kintner worried him**: Frank, *Out of Thin Air*, 202–4.

265 **"This visit to New York"**: Piers Anderton to Birgitta Anderton, n.d. [c. Oct. 1962], courtesy B. Anderton.

266 **"one effect has been"**: Jack Gould, "N.B.C. and Berlin Wall Tunnel," *NYT*, Oct. 22, 1962.

266 **"summoned Producer Wood"**: MGM publicity release, n.d., MPA.

267 **"Dept. may wish contact"**: Allen Lightner to Dean Rusk, Oct. 18, 1962, RFP.

267 **the Stasi already knew**: BStU/MfS, 7413/72.

267 **An early morning meeting**: *PR of JFK*, 2:578–99.

268 **"These brass hats"**: Robert F. Kennedy, *Thirteen Days* (New York: W. W. Norton, 1969).

268 **"The Defense of Berlin"**: David Klein to McGeorge Bundy, memo, Oct. 19, 1962, JFK-NSF.

269 **"You know, the only decent advice"**: Dean Rusk, *As I Saw It* (New York: W. W. Norton, 1990), 243.

269 **His mission as NBC's fixer**: *Variety* clip, Lester Bernstein to Mrs. Bernstein (mother), Oct. 19, 1962, Lester Bernstein Letters, Bernstein family; Frank, *Out of Thin Air*, 204.

270 **"Frankly, the NBC management"**: Lester Bernstein to Egon Bahr, memo, Oct. 19, 1962, Lester Bernstein Letters.

271 **Reuven Frank feared**: Frank, *Out of Thin Air*, 205; Frank, "Making of *The Tunnel*," 22.

271 **A sprawling piece**: Bainbridge, "Die Mauer: The Early Days of the Berlin Wall."

271 **Renowned foreign correspondent**: Flora Lewis, *NYT*, Oct. 28, 1962.

272   **The latest edition**: *Esquire*, November 1962.

273   **"The newsmen roared"**: "Pentagon Issues Shelter Report," *NYT*, Oct. 26, 1962.

273   **The Air Force had ordered**: A copy of this study is in JFK-POF.

274   **The authors had "distorted"**: Jack Raymond, "Pentagon Backs 'Fail-Safe' Setup," *NYT*, Oct. 21, 1962.

274   **premiere of MGM's**: Harry Thoess to Reuven Frank, Jan. 21, 1988, RFP; Frank, *Out of Thin Air*, 209.

276   **Lester Bernstein's family in New York**: N. Bernstein int.

276   **"I believe decisions"**: Adlai Stevenson to Dean Rusk, Oct. 22, 1962, JFK-NSF.

276   **JFK told one of the Joint Chiefs**: *PR of JFK*, 2:102–82.

277   **When he awoke**: Rusk, *As I Saw It*, 235.

278   **When he received the news**: Frank, *Out of Thin Air*, 205–6.

278   **On October 27, a Soviet**: *Boston Globe*, August 7, 2012.

279   **General LeMay**: Robert Dallek, "JFK vs. the Military," *The Atlantic*, Special JFK Issue, Fall 2013.

279   **"Events since then"**: John F. Kennedy to Orvil Dryfoos, Oct. 25, 1962, JFK-POF.

280   ***Der Spiegel* compared it**: *Der Spiegel*, Oct. 31, 1962.

280   **"artistic value"**: *Die Zeit*, Oct. 26, 1962.

281   **"Not long after *Escape*"**: MGM press kit, MPA.

281   **"rather than twenty-eight escapes"**: *BoxOffice*, Nov. 12, 1962.

17. SABOTAGE

282   **More than a year**: Seidel int.; Galante, *Berlin Wall*, 194.

282   **the brothers Franzke**: Boris Franzke int.; Veigel, *Wege durch die Mauer*, 284–88.

284   **afternoon of November 5**: Franzke int.

284   **When Harry told Wagner**: Galante, *Berlin Wall*, 196.

284   **The Franzkes estimated that**: Much of the information in this chapter, not noted separately below, comes from the manuscript for a new chapter in an updated 2015 edition of Veigel, *Wege durch die Mauer*.

285   **Agents had already arrested the Schallers**: Helga (Schaller) Stoof int.

285   **Hoping for more arrests**: Galante, *Berlin Wall*, 196–97; Veigel, *Wege durch die Mauer*, manuscript for new chapter.

286   **the Schallers were away**: Veigel int.

286   **Inside the tunnel**: Seidel, Franzke ints.; Galante, *Berlin Wall*, 198.

287   **The Stasi commander**: Franzke int.

288   **Harry Seidel, in custody**: Seidel int.; also see BStU/MfS, ZKG 7914 5.

288   **Schmeing, defying an order**: Franzke, Veigel ints.

289   **The MfS hastily drafted**: BStU/MfS, HAI I nr. 14633.

290   **Arthur Sylvester set off**: Pierre Salinger, *With Kennedy* (New York: Doubleday, 1966), 285–87.

290   **hue and cry**: For the criticisms of Sylvester, see Coleman, *Fourteenth Day*, 155–56.

290   **Sylvester's views were largely shared**: Salinger, *P.S.*, 125–26; Coleman, *Fourteenth Day*, 157.

291   **Sadly for Sylvester**: "U.S. Aide Defends Lying to Nation," *NYT*, Dec. 7, 1962.

291   **Privately, JFK admitted**: Salinger, *P.S.*, 126–27; Salinger, *With Kennedy*, 293–94.

291   **Beating NBC to the punch**: Review of *Tunnel to Freedom*, *Variety*, Nov. 9, 1962.

291   **submit his resignation**: Frank, *Out of Thin Air*, 206; Frank, "Making of *The Tunnel*," 22–23.

292   **"set the facts before you"**: Robert Kintner to Dean Rusk, Nov. 20, 1962, RFP.

293   **"I understand you will be hearing"**: Pierre Salinger to Robert Kintner, Nov. 29, 1962, Central Files, Box 57, JFKL.

293   **"appease some members"**: Robert Manning to Dean Rusk, Nov. 24, 1962, RFP.

293   **The Rusk letter**: Dean Rusk to Robert Kintner, Nov. 28, 1962, RFP.

294   **Sander Vanocur**: Frank, *Out of Thin Air*, 204.

294   **"That was a terrible thing"**: Ibid., 206.

294   **Uhse's MfS pay had shot up**: BStU/MfS, Uhse file, 13337/64.

294   **"deepen relationship of trust"**: BStU/MfS, Uhse file, Nov. 25, 1962.

295   **Hasso tried to recruit Claus**: I. Stürmer int.

295   **Wolf Schroedter, who still felt guilty**: Schroedter int.

295   **first major magazine piece**: Don Cook, "Digging a Way to Freedom," *Saturday Evening Post*, Dec. 8, 1962.

296   **Echoing some of the reviews**: *NYT*, Dec. 6, 1962; *Los Angeles Examiner*, Dec. 6, 1962; *Variety*, Oct. 23, 1962.

296   **added, real-life attraction**: "German Girl on P.A. Tour to Promote German Film," *BoxOffice*, Dec. 3, 1962.

297   **"I hope I'll be able"**: Lester Bernstein to Robert Manning, Nov. 30, 1962, RFP.

297   **"on the basis of its own judgment"**: NBC press release, Nov. 30, 1962, RFP.

298   **"You handled yourself"**: Robert Kintner to Lester Bernstein, memo, n.d., Lester Bernstein Letters, Bernstein family.

18. COMING UP FOR AIR

299   **Three months into**: H. Stachowitz int.

299   **a visit by her mother**: G. Stachowitz int.

299   **For Manfred Meier**: Meier int.

299   **Harry Seidel also remained**: Galante, *Berlin Wall*, 200–201.

300   **A Stasi interrogator told**: Veigel, *Wege durch die Mauer*, manuscript for new chapter.

300   **The Stasi had probed**: Ibid.; BStU/MfS, ZKG 7914.

301   **The Sendlers were threatening**: BStU/MfS, HAI 13256 47.

305   **The film had not yet aired in Germany**: *Frankfurter Allgemeine Zeitung*, n.d. (c. Dec. 1962).

305   **But watching *The Tunnel***: Bill Moyers int.

306   **Reuven Frank and his team**: Frank, *Out of Thin Air*, 206.

306   **18 million homes**: Ibid.

306   **helped build viewership**: Frank, "Making of *The Tunnel*," 22–23.

307   **Wolf Schroedter's new tunnel project**: Schroedter int.

307   **a different basement**: Herschel, J. Rudolph ints.

308   **bomb blasts**: Alex T. Prengel to Dept. of State, n.d., JFK-BC.

308   **Spirits inside the swizzle stick factory**: J. Rudolph int.

308   **To ease the malaise**: Herschel, J. Rudolph ints.

EPILOGUE

311   **third bomb blast**: *NYT*, Dec. 29, 1962.

311   **Seidel was convicted**: Galante, *Berlin Wall*, 200–206, 263–69.

311   **"have now slipped my memory"**: BStU/MfS 2743/69 240.

312   **Two months after Hesso Herschel's**: Arnold and Kellerhoff, *Fluchttunnel von Berlin*, 241–45.

312   **When he asked the leaders**: Schroedter int.

312   **he seemed "harmless"**: W.-D. Sternheimer int.

312   **Uhse continued his Stasi career**: BStU/MfS, 3733/65 409; BStU/MfS, 13337/64; Veigel, *Wege durch die Mauer*.

312   **Angelika Ligma**: BStU/MfS, 7413/72.

313   **Piers Anderton, still stewing**: B. Anderton int.; "U.S. Censorship Overseas?," *Broadcasting*, Jan. 7, 1963.

313   **"You're my kind of guy"**: Piers Anderton to Birgitta Anderton, Jan. 5, 1963, courtesy of B. Anderton.

314   **"I was called in to the State"**: Ibid.

314   **transferred to India**: B. Anderton int.

314   **one hundred copies**: Frank, *Out of Thin Air*, 207.

314   **criticized the State Department**: Ibid., 206–7.

314   **"for winning it for me"**: Piers Anderton to Birgitta Anderton, May 27, 1963, courtesy of B. Anderton.

314   **"not responsible conduct"**: Jack Gould, *NYT*, June 2, 1963.

315   **When it aired in June 1963**: BStU/MfS, 2743/69.

315   **"So now the record is clear"**: Tim Weiner, "Project Mockingbird: Spying on Reporters," *NYT*, June 26, 2007.

316   **"The networks appeared to"**: Watson, *Expanding Vista*, 144.

317   **Friedrich and Edith Sendler**: BStU/MfS, HAI 13256 47.

317   **In 1964, Joachim Neumann**: Neumann int.; Arnold and Kellerhoff, *Die Fluchttunnel von Berlin*, 251–70; Ahonen, *Death at the Berlin Wall*, 106–27.

318   **Hasso's big Cadillac**: Herschel int.

318   **Manfred Meier's fiancée**: Bayer, Meier ints.

318   **its cruelest tricks**: H. Stachowitz, G. Stachowitz ints.

319   **Harry Seidel's mother**: Seidel int.

319   **"Reuven wrote the book"**: UPI, March 2, 1982.

319 **Frank revealed**: Frank, *Out of Thin Air*, 210–12; Reuven Frank to Dean Rusk, Aug. 8, 1988, RFP; Dean Rusk to Reuven Frank, Aug. 31, 1988, RFP.

320 **"Yes, he certainly was"**: Thomas J. Schoenbaum to author, Nov. 23, 2015.

320 **"My best argument is"**: Reuven Frank interview, "Emmy TV Legends," Archive of American Television, www.emmy legends.org.

320 **Piers Anderton, angry about**: M. Anderton, B. Anderton ints.

321 **"a case of a boss"**: Daniel Schorr interview, Newseum.

321 **a lecture at Harvard**: Theodore H. White Lecture, Shorenstein Center, Harvard University, booklet, 1993, 18–19.

322 **Harry Seidel returned to cycling**: Seidel int.

322 **His former boss**: Ibid.

322 **Hasso Herschel stepped away**: Herschel int.

322 **Joan Glenn**: Ibid.; BStU/MfS, 3733/65 205; BStU/MfS, 18284/63 213; BStU/MfS, 1596/64 175.

323 **Bob Dylan**: BStU/MfS, HA XX, nr. 16578, Bl. 137.

323 **Bruce Springsteen**: Erik Kirschbaum, *Bruce Springsteen: Rocking the Wall* (New York: Berlinica, 2013).

324 **East Berliners were still dying**: Hertle and Nooke, *Victims at the Berlin Wall*, 423–31.

325 **Herschel was cooking**: Herschel int.; Hasso Herschel interview, Newseum.

325 **The same night**: Euronews, March 11, 2014, http://www.euro news.com/.

325 **Rudolph offered to drive**: J. Rudolph int.

326 **weak TV ratings**: *Washington Post*, Dec. 1, 1989.

327 **"Germany's second chance"**: George Packer, "The Quiet German," *The New Yorker*, Dec. 1, 2014.

327 **"Nevertheless, I wish you well"**: *Der Spiegel*, Feb. 1, 1990.

327 **Siegfried Uhse**: Veigel, *Wege durch die Mauer*, 269–71; BStU/MfS, 13337/64.

327 **After the Wall fell**: H. Stachowitz, G. Stachowitz ints.

327 **"You can draw a direct"**: *Ein Tag im August*, Wolfgang Schoen, 2012.

328 **Christian Zobel**: Hertle and Nooke, *Victims at the Berlin Wall*, 453.

328   **Ulrich Mühe**: *The Telegraph* (London), July 27, 2007.

328   **A 2009 poll**: *Der Spiegel*, March 7, 2009.

329   **NBC chose this moment to post**: Article posted Nov. 10, 2014, www.nbcnews.com/Storyline/berlin-wall-anniversary.

329   **Peter Schneider**: Packer, "The Quiet German."

330   **"The wall is a symbol"**: Hare, David, *Berlin Wall* (London: Faber and Faber, 2009), 45.

330   **"left vast empty spaces"**: *NYT,* Nov. 17, 2014.

331   **the black numeral "7"**: J. Rudolph int.

BIBLIOGRAPHY

ARCHIVES

Hanson Baldwin Papers: Yale University, New Haven, Connecticut.

Lester Bernstein Letters (sent by Bernstein to his family and others): Bernstein family.

Der Bundesbeauftragte für die Unterlagen des Staatssicherheitsdienstes (Stasi archives): Berlin, Germany. Note: most documents in the Siegfried Uhse file lack reference numbers in my citations.

Blair Clark Papers: Princeton University, Princeton, New Jersey.

Reuven Frank Papers: Tufts University, Medford, Massachusetts. These were mainly letters Frank sent and received in 1988, while he was writing his memoir, and government documents received during the same period in response to a Freedom of Information Act request made to the State Department. (Many of the latter are also found in JFK-NSF and JFK-BC.)

John F. Kennedy Papers: John F. Kennedy Presidential Library (JFKL), Boston, Massachusetts.

> State Department cables to and from Berlin, Bonn, and Washington, D.C., National Security Files (JFK-BC)
>
> Other National Security Files (JFK-NSF)
>
> Daily CIA briefing, released 2015; see http://www.foia.cia/collection.PDBs (JFK-PDB)
>
> President's Office Files (JFK-POF)

Robert J. Manning Papers: Yale University, New Haven, Connecticut.

Motion Picture Academy: Margaret Herrick Library, Los Angeles, California, files for *Escape from East Berlin/Tunnel 28.*

National Archives and Records Administration, College Park, Maryland.

Berlin Brigade, U.S. military logs, 1962

State Department and CIA documents released in January 2014 by the National Declassification Center; see www.archives.gov/research/foreign-policy/cold-war/berlin-wall-1962-1987/dvd/start.swf (NARA-De)

Declassified CIA internal weekly and periodic intelligence reports (NARA-CIA)

Dean Rusk Papers: Lyndon B. Johnson Presidential Library, Austin, Texas.

Richard S. Salant Papers: New Canaan Public Library, New Canaan, Connecticut.

Daniel Schorr Papers: Library of Congress, Washington, D.C.

FILMS

Berlin Wall Gallery exhibit. Video interviews and transcript of outtakes with Reuven Frank, Daniel Schorr, and Hasso Herschel. Newseum, Washington, D.C.

Ein Tag im August: Der Fall Peter Fechter. Film directed by Wolfgang Schoen. 2012.

Test for the West. Film directed by Franz Baake. 1962–1963.

Der Tunnel. Documentary directed by Marcus Vetter. 1999.

The Tunnel. NBC documentary. Aired Dec. 10, 1962.

INTERVIEWS

Interviews with the following people were conducted by (or, in a few cases, on behalf of) the author in 2015 or 2016. Usually lasting two to four hours or longer, almost all took place in or near Berlin. They are labeled "int." in the notes.

Tunnelers: Harry Seidel, Hasso Herschel, Wolf Schroedter, Joachim Rudolph, Uli Pfeifer, Joachim Neumann, Claus Stürmer, Boris Franzke, Klaus-M. von Keussler.

Escapees: Eveline (Schmidt) Rudolph, Anita Moeller, Gerda Stachowitz, Inge Stürmer, Renate Sternheimer, Britta Bayer, Helga (Schaller) Stoof.

Couriers: Ellen (Schau) Sesta, Hartmut Stachowitz, Wolf-Dieter Sternheimer, Manfred Meier.

At the State Department: James L. Greenfield, Arthur Day.

NBC-related personnel: Abraham Ashkenasi, Birgitta Anderton, Mary Anderton, Kit Anderton, Peter Frank, Jim Frank, Markus Thoess, Renate Stindt, Paul and Nina Bernstein, Sander Vanocur.

At MGM: Christine Kaufmann, Franz Baake.

Experts and others: Burkhart Veigel, Maria Nooke, Dietmar Arnold, Fritjof Meyer, Thomas Bahner, Bill Moyers, Erdman Weyrauch.

ORAL HISTORIES

At JFKL: Elie Abel, Dean Acheson, McGeorge Bundy, Abraham Chayes, Lucius Clay, Walter Lippmann, Dean Rusk, Pierre Salinger, Ted Sorensen, Frank Stanton.

KEY MAGAZINE ARTICLE

Frank, Reuven. "Making of 'The Tunnel,'" *Television Quarterly*, Fall, 1963.

BOOKS

Ahonen, Pertti. *Death at the Berlin Wall*. New York: Oxford University Press, 2011.

Alterman, Eric. *When Presidents Lie*. New York: Viking, 2004.

Arnold, Dietmar, and Sven Felix Kellerhoff. *Die Fluchttunnel von Berlin*. Berlin: Propylaen, 2008. Revised and published as *Unterirdisch in die Freiheit*, 2015.

Barnouw, Erik. *Documentary: A History of the Non-Fiction Film*. New York: Oxford University Press, 1993.

Bluem, A. William. *Documentary in American Television*. New York: Hastings House, 1965.

Bradlee, Benjamin C. *Conversations with Kennedy*. New York: W. W. Norton, 1975.

Coleman, David G. *The Fourteenth Day: JFK and the Aftermath of the Cuban Missile Crisis*. New York: W. W. Norton, 2012.

Dallek, Robert. *An Unfinished Life: John F. Kennedy*. New York: Little, Brown and Company, 2003.

Davies, Robert B. *Baldwin of the "Times."* Annapolis, Md.: Naval Institute Press, 2011.

Detjen, Marion. *Ein Loch in Der Mauer*. Berlin: Siedler Verlag, 2005.

Frank, Reuven. *Out of Thin Air: The Brief Wonderful Life of Network News*. New York: Simon & Schuster, 1991.

Funder, Anna. *Stasiland: Stories from Behind the Berlin Wall*. New York: Perennial, 2002.

Galante, Pierre. *The Berlin Wall*. Garden City, N.Y.: Doubleday, 1963.

Halberstam, David. *The Powers That Be*. New York: Knopf, 1979.

Hertle, Hans-Herman, and Maria Nooke, eds. *The Victims at the Berlin Wall, 1961–1989*. Berlin: Ch Links, 2011.

Hill, T. H. E. *Berlin in Early Wall Era—CIA, State Department, and Army Booklets*. Self-published, 2014.

Hilton, Christopher. *The Wall: The People's Story*. London: Sutton, 2001.

Keil, Lars Broder, and Sven Felix Kellerhoff. *Mord an der Mauer: Der Fall Peter Fechter*. Berlin: Bastei Lubbe, 2012.

Kemp, Anthony. *Escape from Berlin*. London: Boxtree Limited, 1987.

Kempe, Frederick. *Berlin 1961*. New York: G. P. Putnam's Sons, 2011.

Kennedy, Robert F. *Thirteen Days*. New York: W. W. Norton, 1969.

Keussler, Klaus-M. von. *Fluchthelfer: Die Gruppe Um Wolfgang Fuchs*. Berlin: Berlin Story Verlag, 2011.

Kirschbaum, Erik. *Bruce Springsteen: Rocking the Wall*. New York: Berlinica, 2013.

Koehler, John O. *Stasi: The Untold Story.* Boulder, Colo.: Westview Press, 1999.

Leo, Maxim. *Red Love: The Story of an East German Family.* London: Pushkin Press, 2013.

Mann, Ulf. *Tunnelfluchten.* Berlin: Transit Buchverlag, 2005.

Murphy, David, Sergei A. Kondrashev, and George Bailey. *Battleground Berlin: CIA vs. KGB in the Cold War.* New Haven, Conn.: Yale University Press, 1997.

Naftali, Timothy, Philip Zelikow, and Ernest May. *The Presidential Recordings of John F. Kennedy: The Great Crises.* 3 vols. New York: W. W. Norton, 2001.

Nooke, Maria. *Der Veratene Tunnel.* Berlin: Edition Temmen, 2002.

Roy, Susan. *Bomboozled.* New York: Pointed Leaf Press, 2011.

Rusk, Dean. *As I Saw It.* New York: W. W. Norton, 1990.

Salinger, Pierre. *With Kennedy.* New York: Doubleday, 1966.

———. *P.S.* New York: St. Martin's Press, 1995.

Schneider, Peter. *The Wall Jumper.* New York: Pantheon, 1983.

Schoenbaum, Thomas J. *Waging Peace and War.* New York: Simon & Schuster, 1988.

Schorr, Daniel. *Staying Tuned: A Life in Journalism.* New York: Washington Square Press, 2001.

Sesta, Ellen. *Der Tunnel in die Freiheit.* Munich: Ullstein, 2001.

Smyser, W. R. *Kennedy and the Berlin Wall.* Lanham, Md.: Rowman & Littlefield, 2010.

Sorensen, Theodore. *Kennedy.* New York: Harper & Row, 1965.

Taylor, Frederick. *The Berlin Wall.* London: Bloomsbury, 2006.

Veigel, Burkhart. *Wege durch die Mauer: Fluchthilfe und Stasi zwischen Ost und West.* Berlin: Berliner Unterwelten, 2013.

Watson, Mary Ann. *The Expanding Vista: American Television in the Kennedy Years.* New York: Oxford University Press, 1990.

Wyden, Peter. *Wall: The Inside Story of Divided Berlin.* New York: Simon & Schuster, 1989.

PHOTOGRAPH CREDITS

(Clockwise from top left for each page of insert)

PAGE ONE

Harry Seidel: The Granger Collection
Siegfried Uhse: BStU Archives
Heidelberger Strasse: Berliner Unterwelten e.V.

PAGE TWO

Joachim Neumann: Berliner Unterwelten e.V.
Gigi Spina, Mimmo Sesta, and Wolf Schroedter: NBC Universal
    Archives
Hasso Herschel: NBC Universal Archives
Joachim Rudolph: courtesy of Joachim Rudolph

PAGE THREE

Gary Stindt, Reuven Frank, and Piers Anderton: courtesy of Bir-
    gitta Anderton
Anderton at tunnel entrance: courtesy of Birgitta Anderton
Flooded tunnel: NBC Universal Archives

PAGE FOUR

Daniel Schorr: Getty Images
Eveline Schmidt: courtesy of Joachim Rudolph
Dean Rusk and John F. Kennedy: Cecil Stoughton/White House
    Photographs, John F. Kennedy Presidential Library and Mu-
    seum

PAGE FIVE

Sketch of Kiefholz tunnel area: BStU Archives
Stasi agents entering Sendler home: NBC Universal Archives

PAGE SIX

Border guards carrying the body of Peter Fechter: Ullstein Bild
    via Getty Images

PAGE SEVEN

View of Schönholzer Strasse: NBC Universal Archives
Eveline Schmidt: NBC Universal Archives
Ellen Schau: NBC Universal Archives

PAGE EIGHT

Anita Moeller: NBC Universal Archives
Tunnel path: Berliner Unterwelten e.V., courtesy of Boris Franzke
Uli Pfeiffer, Joachim Rudolph, Joachim Neumann, and Hasso
    Herschel: Berliner Unterwelten e.V., courtesy of Boris
    Franzke

INDEX

ABOUT THE AUTHOR

GREG MITCHELL is the author of a dozen books, including *The Campaign of the Century*, winner of the Goldsmith Book Prize and finalist for the Los Angeles Times Book Prize, on Upton Sinclair's race for governor of California in 1934 and the birth of media politics; *Tricky Dick and the Pink Lady*, a New York Times Notable Book; *So Wrong for So Long: How the Press and the President Failed on Iraq*; and, with Robert Jay Lifton, *Hiroshima in America* and *Who Owns Death?*

Mitchell has edited several national magazines, including, from 2001 to 2009, *Editor & Publisher*, where he won numerous awards for media coverage. He recently coproduced a documentary, *Following the Ninth*, about the global influence of Beethoven's famous symphony (including its role in Berlin when the Wall fell), and coauthored a book with the film's director, Kerry Candaele. Earlier he served as a chief consultant to two other award-winning documentaries, *Original Child Bomb* and *The Great Depression*.

Mitchell's articles have appeared in dozens of leading magazines and newspapers. He blogs regularly for *Huffington Post* and other outlets. His Twitter feed is @GregMitch, and his personal blog is *Pressing Issues*. He lives in the New York City area.